Abuse and Neglect of the Elderly in India

Mala Kapur Shankardass
S. Irudaya Rajan
Editors

Abuse and Neglect of the Elderly in India

Springer

Editors
Mala Kapur Shankardass
Maitreyi College
University of Delhi
New Delhi
India

S. Irudaya Rajan
Centre for Development Studies
Thiruvananthapuram, Kerala
India

ISBN 978-981-10-6115-8 ISBN 978-981-10-6116-5 (eBook)
https://doi.org/10.1007/978-981-10-6116-5

Library of Congress Control Number: 2017948234

Printed on acid-free paper

This Springer imprint is published by Springer Nature
The registered company is Springer Nature Singapore Pte Ltd.
The registered company address is: 152 Beach Road, #21-01/04 Gateway East, Singapore 189721, Singapore

Foreword

Abuse and neglect of the elderly is an emerging concern across many countries, which have grave human rights implications. The book highlights different aspects of this problem, suggesting means of preventing the social ills. The book uses data from a number of population-based studies to describe the different dimensions of elder abuse and neglect among different groups and sections in society. The book also provides a background of socio-demographic characteristics to help understand some of the key reasons that perpetuate this problem. It is a very relevant contribution to understand the context in which elder abuse and neglect in India occurs and it also contrasts the knowledge with global experiences. The book covers current perspectives on elder abuse and neglect, at the household level, among widows and among those residing in institutions. It covers rural and urban areas in different states of India, some of which have populations with a high proportion of older people. One of the relevant features of the book is presenting data which has not been discussed in such detail before, which will be useful for policy and programming for the well-being of the elderly population in India.

I am sure that this book will be of immense interest to researchers, academicians, policy makers, civil society members, universities, research institutes, elder people and the public at large. The book will also serve as a good reference material for the courses in social gerontology, sociology, social work at graduation and post-graduation and higher levels at universities.

I place on record my sincere appreciation to the editors of this book for their valuable contribution.

New Delhi, India
Feburary 2017

Diego Palacios
Representative UNFPA India

Acknowledgements

The idea for this volume arose from the success of the sessions on elder abuse and neglect at the 17th Biennial Conference of Association of Gerontology, India and the International Conference on Engaging and Empowering the Elderly held at the Centre for Development Studies in Thiruvananthapuram, Kerala, India (15–16 September 2014). The then President of the Association of Gerontology, India, Prof. S. Irudaya Rajan (one of this book's editors) very encouragingly put forward the idea of this publication to the other editor (Dr. Mala Kapur Shankardass, who holds the Asia Chair of the International Network for Prevention of Elder Abuse and was a participant in the elder abuse sessions of the conference) to jointly approach the subject in book form through the works of experts in the field. Some of these experts had participated in the sessions on abuse at the above-stated conference, and others who had long worked on the topic. We are thankful that with our joint efforts we could conceptualize the issue and bring forth an intellectual discourse on the topic which has emerged as a growing concern requiring theoretical understanding and empirical deliberations not only in India but also abroad.

As editors, we would like to thank all the contributors to this volume for sharing their ideas and insights in this book. We are extremely grateful to some of our friends and colleagues in India and a few from abroad, all experts in the field of elder abuse and neglect, who agreed without any hesitation or questioning to contribute to this special publication. We are indebted to all of them: A.M. Khan, Ajay Bailey, Anindita Datta, Anupama Datta, Asha Banu Soletti, Ashish Goel, Astha Koolwal, D. Jamuna, Daliya Sebastian, Muthuvenkatachalam Srinivasan, Nidhi Gupta, Nishesh Jain, P.N.N. Nikhil, Pankhuri Bhatia, Renu A. Varughese, Ruchika Kuba, S. Siva Raju, Sangeeta Kumari Gupta, Sandhya Gupta, Selim Jahangir, T.V. Sekher and U.R. Arya. We are also grateful to Diego Palacios, Representative, UNFPA India for writing the Preface and extending his support.

It has been a pleasure working on this volume with Shinijini Chatterji, editor at Springer. She has guided and nurtured this volume from its beginning, when she willingly accepted the concept, right through to its completion and publication.

We would also like to thank Praveen Kumar and Priya Vyas from Springer for their association and assistance with the book.

We are grateful for the opportunity to bring out this book and hope it will be of interest to a wide section of the population including professionals, researchers, practitioners, policymakers and the lay public, and particularly those interested in elder abuse and neglect.

Finally, we would like to thank our families for their support and cooperation while we worked on this book.

Dr. Mala Kapur Shankardass
Dr. S. Irudaya Rajan

Contents

Editors and Contributors

About the Editors

Mala Kapur Shankardass is Associate Professor at Maitreyi College, University of Delhi (South Campus), India, and holds a doctoral specialization in Medical Sociology and post-doctorate in ageing studies. In addition, she holds a diploma in Gerontology from the International Institute of Ageing (INIA), Malta, and in Demographic Aspects of Population Ageing. She is a Fellow of the World Demographic Association.

She was a panelist at the United Nations (UN) General Assembly on the International Conference on Population and Development (ICPD) and at the Second World Assembly on Ageing, where a UN Fund for Population Activities (UNFPA) book was released with her chapter on 'Voices of Older People in India'. She was part of the Technical Expert Group for the International Plan on Ageing and was involved in developing the Research Agenda on Ageing for the UN Programme on Ageing. She was invited by the Spanish government to review the implementation of the Madrid International Plan on Ageing in 2005. From 2004 to 2007 she was a consultant with the UN Population Fund Country Support Team (UNFPA CST) for the project 'Ageing in Asia'. In 2007 she developed a Strategy Paper on Ageing for the UNFPA, and in 2008 she served as an advisor to the World Health Organization South East Asia Regional Office's Healthy Ageing programme. Dr. Shankardass has served as a consultant to the International Institute of Ageing Studies, Malta; Colombo Plan, Sri Lanka; and Asia Training Centre on Ageing, Thailand. She was involved with the formulation of the National Policy for Older Persons in India, and is a member of the Core Group on Protection and Welfare of Elderly Persons, created by the National Human Rights Commission.

She publishes in reputed journals, magazines, newspapers and books. She authored the book *Growing Old in India: Voices Reveal and Statistics Speak*. She is Asia representative of the International Network for Prevention of Elder Abuse (INPEA) and Member of the Editorial Board for the *Journal of Adult Protection* and the *Polish Social Gerontology Journal*.

S. Irudaya Rajan is Professor at the Centre for Development Studies, Thiruvananthapuram, Kerala. Currently, he is the President of the Association of Gerontology (AGI, India) as well as Kerala Economic Association (KEA). He has three decades of research experience and has published extensively in national and international journals on social, economic, health and demographic facets of population ageing. He is the lead author of the book *India's Elderly: Burden or Challenge?* (1999) and has edited the books *Social Security for the Elderly: Experiences from South Asia* (2008) and *Elderly Care in India: Societal and State Responses* (Springer, 2017). He developed the first longitudinal Ageing Survey in Kerala (India) in 2004; repeated every three years, it completed its fifth wave in 2016 and will enter its sixth in 2019. The second longitudinal Ageing Survey in Kerala started in 2013, completed its second wave in 2016 and will enter its third wave in 2019.

Professor Rajan has been involved in several projects on ageing funded by the UN, UN Fund for Population Activities (UNFPA), World Bank (WB), International Labour Organization (ILO), HelpAge International, South Asian Network of Economic Institutes, Indo-Dutch Programme on Alternatives in Development (IDPAD), Shastri Indo-Canadian Institute and International Development Research Centre. He has also undertaken considerable research on international migration and is the editor of two series—India Migration Report and South Asia Migration report—as well as being Editor-in-Chief of the journal *Migration and Development.*

Contributors

U.R. Arya is a Ph.D. scholar and doctoral fellow at the Centre for Development Studies, Thiruvananthapuram whose current research area is active ageing in Kerala.

Dr. Ajay Bailey is an Assistant Professor at the Population Research Centre at the Faculty of Spatial Sciences in the University of Groningen. He is also the India research and training coordinator for the Faculty of Spatial Sciences. He leads the research line 'Global Migration, Culture and Place'. He is an anthropologist and a cultural demographer by training. His main research interests include migration, culture, access to health-care and mixed methods. He has received seven grants for research and research networking/training. For more information: https://www.rug.nl/staff/a.bailey.

Nasir Ahmad Bhat is a Ph.D. scholar in the Department of Psychology at Jamia Millia Islamia, New Delhi.

Pankhuri Bhatia graduated from Lady Shri Ram College, Delhi in 2005 and over the next four years completed her Masters in English Literature and M.Phil. in English Language Teaching. After teaching for about a year, she went on to do her second Masters from Tata Institute of Social Sciences (TISS), Mumbai. In 2016, she graduated from TISS, Mumbai with a Master's in Social Work (specialization in mental health). At TISS, she was awarded as the best student in M.A. in social work (Mental Health) and was also adjudged as the best student in Field Work. She went to Gothenburg University, Sweden as part of a Student Exchange programme and passed with distinction there.

While in TISS, she researched on Partner Abuse in Later Life. She is currently working as a Research Associate in the Centre on Death Penalty, National Law University, Delhi. She is involved in a research project on the mental health status of prisoners sentenced to death in India.

Anindita Datta is Associate Professor at the Department of Geography, Delhi School of Economics. Her research interests are broadly in the area of feminist geographies and the Geographies of Gender. She has published consistently in internationally known peer-reviewed journals with an inter-disciplinary perspective, and is on the international editorial board for *Gender Place and Culture*. She is a member of the steering committee of the International Geographical Union's Commission for Gender and Geography.

Anupama Datta has worked with HelpAge India for the last 15 years and has headed the Policy Research and Development Department for the last 10 years. The department deals with research and advocacy for rights of older persons in India. The main research projects undertaken were on elder abuse, pensions, disasters and older persons. Advocacy initiatives (in collaboration with the Ministry of Social Justice and Empowerment, NHRC and various international organizations) include organizing seminars and discussions on various topics concerning ageing and elderly people in India. She has published reader-friendly booklets on various topics (such as preventive and curative aspects of health issues like diabetes, dementia, cardiovascular diseases, and other topics like universal design, etc.). Her other publications include information-based booklets like the *Senior Citizens' Guide*. She is the Editor of *HelpAge India Research and Development Journal*, published three times a year.

Ashish Goel is an Associate Professor in the Department of Medicine at the University College of Medical Sciences at the University of Delhi. He completed his graduation from Mahatma Gandhi Institute of Medical Sciences, the first rural medical college in India, and subsequently trained at the All India Institute of Medical Sciences. Groomed in a Gandhian atmosphere at Sevagram, he believes in simple living and high thinking. His academic interests include needs of special populations in disaster situations, end-of-life care issues, care of frail older people, medical ethics and medical education. He enjoys cooking, reading, travel and photography. He has over 100 scientific publications in national and international journals to his credit. He has contributed nine books and numerous other book chapters. He has organized and addressed several national and international conferences.

Nidhi Gupta is a versatile public health professional and works as faculty at the Centre for Population, Health and Development, School of Development Studies, Tata Institute of Social Sciences, (TISS), Mumbai. She is coordinating the United Nations Population Fund project on ageing in India, at TISS. Besides her active engagement in teaching, health and social research, she has rich clinical experience. She visited Norwegian University of Science and Technology (NTNU), Trondheim to conduct a research study on models of social and healthcare for older adults in Norway. She has presented/published several research papers on issues related to ageing and health, reproductive health and health systems in various international and national conferences/peer-reviewed journals. She is a reviewer for various peer-reviewed journals. She is a recipient of various fellowships and awards such as Linnaeus-Palme Exchange International Exchange Programme at Karolinska Institutet, Sweden and UGC-SRF, Government of India during doctoral work, Population Association of America (PAA), travel award to present her paper at the 2016 Annual Meeting, Washington. Her research interests include ageing and health, quality of life, feminization of ageing, health and social care models for elderly people, primary healthcare, technology and ageing, quality of care, and gender issues.

Sandhya Gupta has a Ph.D. (Geriatric Care Services) and M.Sc. (Psychiatric Nursing), and has worked as faculty at the College of Nursing, AIIMS, New Delhi for 36 years. She is a Ph.D. guide. She is a master trainer in counselling, in HIV/AIDS, for substance use disorders and for mental health nursing. She has been decorated with the President's gold medal, Prashasthi award, WHO fellowship to Australia, Fellow of Indian society of psychiatric nurses and Best Research Work Award of AIIMSONIANS of America. She has the honour of being a member of the expert advisory committee for the National Mental Health Programme, Mental Health Bill, on Old-Age Homes. She has written books on substance use disorders, modules for care of older persons, counselling for HIV care, psychosocial care and disaster preparedness; and written chapters on geriatric nursing, palliative care for elderly, etc. She is Chief Editor of the *Indian Journal of Psychiatric Nursing*. She has had more than 55 articles in scientific journals.

Selim Jahangir is a doctoral student in the Department of Geography, University of Delhi. He has an M.Phil. in Geography from the Department of Geography, Delhi School of Economics, University of Delhi. His doctoral works focus on 'Gender, Space and Ageing: Care Geographies of Elderly Men in Delhi and Kolkata'. His research interests are in the varied fields of social geography and social gerontology.

Nishesh Jain gained his doctorate in Internal Medicine. He is a Senior Fellow in Diabetology and Metabolism, Member of American Association of Clinical Endocrinology, Indian Thyroid Society, Research Society of Study of Diabetes in India, American Thyroid Society. He has done a postgraduate course in Diabetology from the prestigious Boston University (School of Medicine). He participated in the practical Diabetology Programme by Cleveland Clinic, Cardiff University and online programme by American Diabetes Association. He participated in clinical programmes by the Indian Thyroid Society and American Thyroid Society. He has contributed to national and international journals on diabetes, especially on the elderly population including papers on elder abuse. He is currently working on several national projects on iodine deficiency in adolescents and society, young and old diabetes-related micro- and macrovascular complications. He also currently works in the prestigious Diabetes Endocrine and Metabolic Centre of University College of Medical Sciences.

D. Jamuna is Professor in the Department of Psychology at Sri Venkateswara University with a Ph.D. in Psychology of Ageing (1984). In addition to a Master's degree in Psychology, she holds Master's Degrees in Education, Women's Studies and a Post Master Diploma in Guidance and Counselling. She was trained in Social Gerontology at the UN International Institute on Ageing, Malta (Europe) on an UNFPA Fellowship (1992); was recipient of Sandoz International Training Fellowship Award of SFGR, Basel, Switzerland to work at the Gerontology Center, Penn State University, USA (1996–1997); was a visiting scholar at the Tokyo Metropolitan Institute of Gerontology on Government of Japan Fellowship (1997–1998); was a recipient of the Meritorious Teacher Award for Research on Ageing of S. V. University (2000). More recently (2003–2004) she was awarded a Fulbright Senior Fellowship Washington to work at the Gerontology Center, Penn State University (PA) and at the Andrus Gerontology Center, University of Southern California, Los Angeles (CA). She has directed 13 national and international research projects; published 84 research articles and edited three books on Gerontology. She acted as President of Association of Gerontology India (AGI) (2006–2008), the professional body of gerontologists. She served as Head of the Department of Psychology and is currently chairperson of Board of Studies (BOS) in Psychology.

A.M. Khan retired in the year 2013 as Professor and Head of Social Science from the National Institute of Health and Family Welfare (NIHFW). He has work experience of 10 years as Director of Two Months Training Course on Child Development for Project Officers under ICDS program of Central Government. He has also organized around 15 national-level seminars, workshops and conferences. As course coordinator and co-coordinator, he has conducted one-month training course on Health System Research (HSR) for senior professionals from universities and medical colleges. During the service, he guided 20 theses and 30 research projects, and published more than 100 articles in national and international journals, books, compendiums, research reports, and proceedings of seminars and conferences. He has visited many countries, gaining short and long-term exposure and experience. He completed a Training Certificate in Social Gerontology from Malta under a UN Fellowship, and worked as Director of Astron Institute of Social Service for about one year. He received an ICSS Senior Fellowship awarded in 2015. He serves as an expert in several national-level organizations and provides special services in the area of research methodology and ageing. He has also served as co-chairperson of Hippocrates Independent Ethics Committee, as a Member of the Board of Studies, as a Member of the Ethics Committee AIMIS, as an Executive Member of the Indian Academy of Social Sciences, and as a Member of the National Committee of Cell Research in ICMR.

Astha Koolwal completed her MBBS from Government Theni Medical College, Theni, Tamil Nadu. She graduated in 2013. She has undergone multiple training rotations in different universities of the USA, namely: Mount Sinai School of Medicine, Manhattan, Thomas Jefferson University, Philadelphia, Children's Hospital of Pittsburgh, Pennsylvania, and Cooper University in New Jersey. In the past she has been involved in numerous research projects, a recent one in Cooper University, New Jersey was based on Hypoxemic Ischemic Encephalopathy in neonates. Currently, she is a research assistant and medical officer at the Anti-Retroviral Centre in the University of Medical Sciences, New Delhi.

Ruchika Kuba is an MBBS, MD (PSM) from LHMC, New Delhi, India. She is currently working as Associate Professor in the School of Health Sciences, Indira Gandhi National Open University, where she has been since 1994. She has coordinated planning, development and maintenance of various postdoctoral, PG and Certificate programmes in geriatric, dialysis, patient safety, health-care waste, cardiology, etc. for health professionals. These include projects funded by WHO and the Ministry of Health. She has been involved as a geriatric expert by other institutes such as the Institute of Home Economics, University of Delhi, National Institute of Social Defence, the Ministry of Social Justice and Empowerment; and for the Government of India in various capacities as a paper setter, member of interview board, external faculty and examiner. She has developed various videos related to geriatrics, and broadcast talks on the general overseas services for All India Radio to more than 100 English-speaking countries. She has authored chapters in textbooks, published papers in indexed journals, and chaired sessions and presented papers in various national and international conferences.

Sangeeta Kumari Gupta is Senior Project Officer (Research) in 'Longitudinal Ageing Study in India (LASI)' at the International Institute for Population Sciences (IIPS), Mumbai. She has a doctorate in Population Studies from the International Institute for Population Sciences, focusing on elderly widows. She has done extensive research and analysis in the area of ageing, health and population. She has also published articles on ageing in various national and international journals. Her book on *Elderly Widows in India* is (2017)'s Elderly Widows in India." –> from Rawat Publications, Jaipur. She has participated in the workshop on Social Gerontology at the International Institute on Ageing, (United Nations), Malta.

P.N.N. Nikhil is a doctoral student at the Population Research Centre, University of Groningen and the Institute for Social and Economic Change, Bangalore. He has a Master's in Population Sciences from the International Institute for Population Sciences (IIPS), Mumbai, India. His doctoral research examines the aspects of 'home-making', place attachment and belongingness of older adults in institutional care settings both in Kerala, India.

S. Siva Raju is Deputy Director, Tata Institute of Social Sciences (Hyderabad Campus). His broad fields of interest for research projects are ageing, health and development. His major publications include: *Regional Development and Family Planning*; *Classified Bibliography on Social Gerontology in India*; *Gerontological Social Work in India*; *Health Status of the Urban Elderly; Adolescents Reproductive Health Perspectives*; *Voice of the Elderly in India*; *Primary Health Centers in Rural Health*; *Social Mapping and Networking of Organizations Working for Care of Older Persons*; *Social Inclusion and Women Health;* and *Ageing, Health and Development and Implementing Corporate Social Responsibility: Indian Perspectives.* He holds a Status Report on older poor people in India, which was launched at the Second World Assembly on Ageing by the United Nations. He is a member of the Committee on Protection and Welfare of the Elderly Persons for the National Human Rights Commission; Expert Committee Member on Ageing for the Ministry of Social Justice and Empowerment; Honorary

Director of International Longevity Centre, Pune; Advisor to the Ministry of Social Security, Government of Mauritius; and a Coordinator of the UNFPA Initiative on Building Knowledge-Base on Population Ageing in India.

Daliya Sebastian is a Research Specialist at Child Fund India, New Delhi. She has a doctorate in Population Studies from the International Institute of Population Sciences (IIPS), Mumbai. Her doctoral work on elder abuse in Kerala was well appreciated by the researchers and also received an ICSSR publication award. As a statistician and demographer, she has authored many journal articles and chapters in edited books on ageing and related issues. Her book *Elder Abuse in India* (Rawat Publications, Jaipur) was a very insightful study on ill-treatment of the aged population, both in household and institutional settings.

T.V. Sekher is Professor at the International Institute for Population Sciences (IIPS), Mumbai. Trained in Demography and Sociology, his areas of research interest are social demography, gender issues, public health and population ageing. He was a Visiting Fellow at the Wellcome Trust Centre of University College London, Maison des sciences de l'homme, Paris, Oxford Brookes University, UK and Lund University in Sweden. He also served as a consultant to UNFPA and UNICEF. Currently, he is one of the national coordinators of the Study on Global Ageing and Adult Health-SAGE India (sponsored by WHO), and Longitudinal Ageing Study in India-LASI (IIPS with Harvard School of Public Health and University of Southern California). He has authored and edited 10 books and has more than 70 research articles in national and international journals.

Asha Banu Soletti is currently the Professor and Chairperson at the Centre for Health and Mental Health, School of Social Work, Tata Institute of Social Sciences, Deonar, Mumbai-88, and teaches Master's level courses for the thematic areas of Social Work in Public Health and Social Work in Mental Health. Areas of research include children affected by HIV/AIDS, ageing and health, homeless and tribal elderly people. Her field of practice and core expertise include community health, mental health, community mental health and development, HIV/AIDS, gerontology, cancer care and palliative care. She has extensive field engagement both in the urban and rural context through her field action projects and other community-level interventions. Presently, she coordinates a field action project 'Integrated Rural Health and Development Project' which is located in a tribal belt and caters to the health needs of the Konkana, Warli and Katkari tribes through a social determinants framework. She has actively responded and contributed to the disaster relief work operations of the Institute. She is also actively involved in training and capacity building of diverse stakeholders in her field of practice.

Muthuvenkatachalam Srinivasan is a postgraduate in Psychiatric Nursing from the All India Institute of Medical Sciences (AIIMS), New Delhi and a Ph.D. scholar from the National Consortium for Ph.D. in Nursing of INC and RGUHS, Bangalore. He is teaching graduate and postgraduate courses in Nursing. He has published more than 10 research papers in national and international journals and presented various papers in national and international conferences. He was awarded an ICMR grant for presenting a research paper at an international conference in Miami, Florida, USA. He is an editorial board member of the *Journal of Psychiatric Nursing*. He received a Sarla Kapur best researcher award from ISPN in February 2016.

Renu A. Varughese is Fulbright Fellow (2015–2016), Mercy College, New York, (USA), currently affiliated to the Centre for Research on Ageing, Department of Psychology, S.V. University, Tirupati, India.

Chapter 1
Introduction

Mala Kapur Shankardass and S. Irudaya Rajan

The genesis of this volume was the 17th Biennial Conference of Association of Gerontology, India and the International Conference on Engaging and Empowering the Elderly held at the Centre for Development Studies in Thiruvananthapuram, Kerala, India in September 2014. At this meeting many papers on the topic of abuse and neglect of the elderly were presented, indicating the popularity of the theme and the need to pay more theoretical and empirical attention to the topic. The editors, both involved with the conference (Shankardass as a participant and session chair on elder abuse, Rajan as the organizer in his capacity as the President of the Association of Gerontology), joined together afterwards and decided to publish a volume by inviting contributions from a few of the conference presenters and others whose work has aided better understanding on the topic.

I

We start the chapter by presenting some socio-demographic data on ageing derived by Rajan from the latest census of India, conducted in 2011, and the Kerala Ageing Survey, 2013, conducted in the Indian state with the largest proportion of older people, which needs to be considered and kept in mind as we reflect on the problem of abuse and neglect of the elderly in India. The rapidly growing population of older people, with state variations; the increasing proportion of older women compared to

Mala Kapur Shankardass (✉)
Department of Sociology, Maitreyi College, South Campus, University of Delhi, New Delhi, India
e-mail: mkshankardass@gmail.com

S. Irudaya Rajan
Centre for Development Studies, Trivandrum, Kerala, India
e-mail: rajancds@gmail.com

Mala Kapur Shankardass and S. Irudaya Rajan (eds.), *Abuse and Neglect of the Elderly in India*, https://doi.org/10.1007/978-981-10-6116-5_1

older men; more widows than widowers, with high levels of illiteracy; labour force participation; living arrangements; migration of children, leaving older people alone and lonely, with different types of disabilities and morbidities; failing mental health status—all these issues set the tone for understanding the growing problem of abuse and neglect of the elderly in India.

The change in people's fertility behaviour and longer life expectancy whether in a healthy or morbid condition affects the age structure of the population. The trend in the elderly population is for their numbers to increase upwards as the younger population is heading downwards. India experienced a situation as an 'ageing nation' in 2001, when the proportion of elderly crossed 7% of India's population of 1.02 billion. According to the 2011 Census, 8.6% (104.2 million) of the total population constituted the elderly in India (Rajan and Sunitha 2015; see Fig. 1.1). After 2021, there will be a tremendous increase in the number of elderly persons.

This will be true especially in the case of elderly people aged 70 and above. As a proportion of the population they are expected to increase by 3.2 times from 2011 to 2051, and by 6.8 times by 2101. Similarly, the proportion of those aged 80 or above is expected to increase by 2.7 times in 2051 and 8.2 times by 2101 (Fig. 1.2).

In 1961 the proportion of elderly was just 5.6% and after 100 years, it will reach 20.2% in 2061. By the end of the century, it is expected that a percentage of elderly persons as high as 28.4% will be surviving in India.

Kerala has the highest proportion of elderly people (12.6%), and Dadra and Nagar Haveli has the lowest proportion (4.0%). Rural areas have more elderly people, especially female elderly ones, compared to urban areas.

In 2011, Uttar Pradesh led with the highest number of elderly people (15.4 million), followed by Maharashtra (11.1 million), West Bengal (7.7 million) and Tamil Nadu (7.5 million).

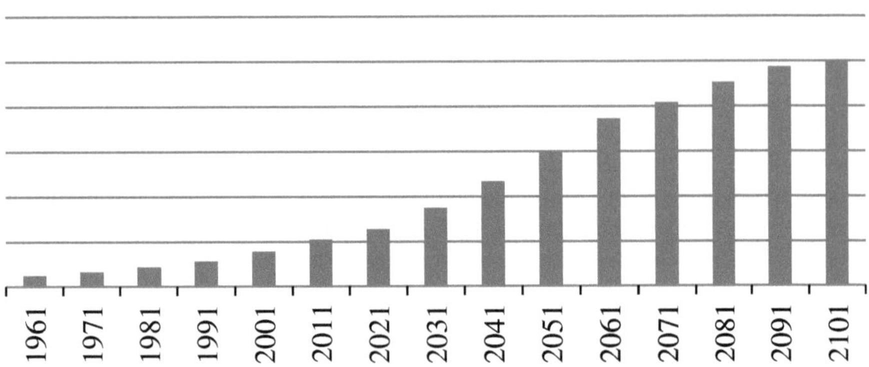

Fig. 1.1 Number of elderly people (in millions). *Source* Compiled by the author from Census 1961–2001, Registrar General of India. *Note* Projected by the author based on Census 2011

Growth Rate

The transition in the population age structure results in a progressive increase in relative and absolute numbers of aged people in India. The aged population has grown steadily from 1971 to 2011. After that there will be a steep decline towards the year 2021 and an increase towards 2031. In future the average annual exponential growth rate of the elderly is expected to decrease from 3.1% in 2031 to 0.3% in 2101. This decline could be in line with the reduction in the current fertility rate. Though there will be an increase in the absolute number of elderly people over the years, the intercensal growth rate will reduce (Fig. 1.3).

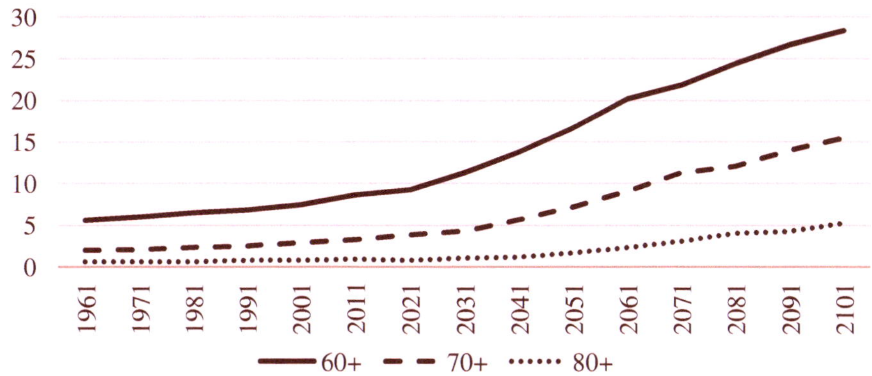

Fig. 1.2 Proportion of elderly people in India, 1961–2101 *Source* Compiled by the author from Census 1961–2001, Registrar General of India

Fig. 1.3 Exponential growth rate of elderly people in India, 1960–2101

Current Scenario of the Elderly

There are 24 'ageing states and union territories' in India according to the Census 2011. These are (with numbers in millions): Jammu and Kashmir (7.4), Himachal Pradesh (10.2), Punjab (10.3), Uttarakhand (8.9), Haryana (8.7), Rajasthan (7.5), Uttar Pradesh (7.7), Bihar (7.4), Manipur (7.0), Tripura (7.9), West Bengal (8.5), Jharkhand (7.1), Odisha (9.5), Chhattisgarh (7.8), Madhya Pradesh (7.9), Gujarat (7.9), Maharashtra (9.9), Andhra Pradesh (9.8), Karnataka (9.5), Goa (11.2), Lakshadweep (8.2), Kerala (12.6), Tamil Nadu (10.4) and Puducherry (9.7). In terms of the proportion of elderly people in the population, Kerala ranks first, with Goa occupying the second largest position followed by Tamil Nadu, Punjab and Himachal Pradesh.

The proportion of elderly people in 11 states seemed to be high in urban areas compared to rural areas. In Kerala, this proportion was the same for both rural and urban areas. The highest proportion can be seen among the southern parts of India such as Kerala, Tamil Nadu, Andhra Pradesh, Karnataka and Goa. Kerala had the highest proportion of very old people (aged 80+) followed by Himachal Pradesh and Punjab, and was concentrated in rural areas (see Table 1.1).

Elderly People by Broad Age Group (Projected)

Elderly people are categorized into three age groups according to their longevity; young old (aged 60–69), old old (aged 70–79) and very old (aged 80+). Normally, young old form the highest proportion of the elderly. But we can see a gradual decrease in the proportion of this younger group and a corresponding increase in the other two age groups towards the end of this century. In the young old category, the proportion of men is high compared to women, but in the other two age groups there are more women. Thus, it can be specifically stated that female elderly persons are living than their male counterparts (see Table 1.2).

Table 1.1 Current age structure of elderly population by sex and place of residence, 2011

	India	60+	70+	80+
Total	Total	8.6	3.3	0.9
	Male	8.2	3.1	0.8
	Female	9.0	3.5	1.0
Rural	Total	8.8	3.4	0.9
	Male	8.4	3.2	0.9
	Female	9.2	3.5	1.0
Urban	Total	8.1	3.1	0.9
	Male	7.7	2.8	0.8
	Female	8.5	3.3	1.0

Compared to 2011 figures, there will be a 14% decrease in the proportion of males among the young old by 2101, while an 8% increase will occur in the other two age groups of the male elderly. The situation of the female elderly is not different except in its percentage level. The decrease in the proportion of young old female elderly persons between 2011 and 2101 is expected to be 18% while a 9% increase in other two age groups will occur.

Sex Ratio

Like other parts of the world, India has an elderly sex ratio favourable to females. Since the female life expectancy is higher at older ages, women comprise a significant part of the elderly population. In Fig. 1.4, there is a fluctuation in the sex ratio year by year but the female population is expected to be high in 2051. In other words, by 2051, there will be only 896 male elderly people for every 1000 female ones. The sex ratio is expected to have peaked this century in 2001.

Table 1.2 Percentage distribution of elderly people by broad age group, 2011–2101

	60–69			70–79			80+		
	Total	Male	Female	Total	Male	Female	Total	Male	Female
2011	61.7	62.0	61.5	27.4	27.7	27.1	10.9	10.3	11.4
2021	58.2	59.9	56.6	33.1	31.9	34.2	8.7	8.2	9.2
2031	61.8	63.5	60.3	29.0	28.5	29.5	9.2	8.0	10.2
2041	59.0	60.1	58.0	32.2	32.1	32.4	8.7	7.8	9.6
2051	56.9	59.2	54.8	32.8	31.8	33.8	10.3	9.1	11.4
2061	55.1	58.0	52.5	33.4	32.4	34.4	11.5	9.7	13.2
2071	48.2	50.2	46.4	37.6	37.6	37.7	14.2	12.2	16.0
2081	50.6	52.7	48.8	32.8	32.5	33.0	16.6	14.8	18.2
2091	47.7	49.7	45.8	36.3	36.0	36.6	16.0	14.3	17.6
2101	45.7	47.6	43.9	35.8	35.6	36.0	18.5	16.8	20.0

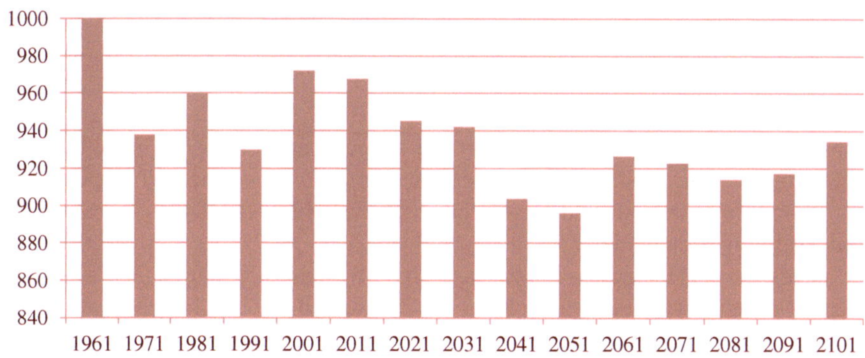

Fig. 1.4 Sex ratio (M/F × 1000) among the elderly in India, 1961–2101

Table 1.3 Elderly people by marital status, 2011

	India	Never married			Currently married			Widowed		
		P	M	F	P	M	F	P	M	F
Total	60+	2.5	2.9	2.0	65.6	82.1	49.6	31.5	14.6	47.8
	70+	3.0	3.7	2.4	53.4	73.8	33.9	43.1	22.1	63.2
	80+	5.9	7.5	4.5	42.9	62.0	26.1	50.8	30.1	69.0
Rural	60+	2.2	2.7	1.7	65.7	81.5	50.4	31.7	15.4	47.4
	70+	2.7	3.4	2.0	53.6	73.4	34.3	43.3	22.9	63.2
	80+	5.3	6.8	4.0	43.3	61.9	26.3	51.0	30.9	69.4
Urban	60+	3.1	3.3	2.9	65.4	83.6	47.7	31.0	12.7	48.8
	70+	3.9	4.4	3.4	52.9	75.0	32.7	42.8	20.3	63.4
	80+	7.3	9.4	5.6	42.0	62.1	25.7	50.3	28.1	68.3

P means persons

Marital Status

Among the total number of elderly persons, 3% were unmarried, and there were more male elderly persons than female ones (Table 1.3). This is true among the currently married elderly. The incidence of widowhood was very high among female elderly people irrespective of their place of residence, but the percentage will increase in line with the increase of age for both sexes. As they get older, elderly females, due to their increasing life expectancy, are always likely to outnumber males as surviving spouses. The percentage of widows were greatest in Tripura (60.5%) followed by Kerala, and Daman and Diu. About 84% of the widows in Kerala are aged above 80 years particularly in urban areas, which have the highest concentration in India followed by Tripura, Daman and Diu Puducherry. The frequency of widowhood among elderly people has been increasing according to their age. Elderly widows are more susceptible to abuse and neglect and, in some cultures, to 'property grabbing' and abandonment (WHO/INPEA 2002). In terms of socio-economic status as well, elderly widows are severely lagging behind and many of them are susceptible to crippling states of poverty (WHO/INPEA 2002; Sethi et al. 2011).

Living Arrangements

The living arrangements of the elderly are a significant dimension reflecting their well-being at their dusk. Rapid changes in the socio-economic and demographic circumstances have reshaped the structure of the family leading to the emerging form of living arrangements.

As per the 2011 Census, the average size of Indian households is 4.5 and the average number of elderly people in the household is 2.4. In other words, an average Indian family has more than two elderly persons to look after. Of the 248.8

million households in India, 31.3% have at least one elderly person, which is higher in rural areas (32.5%) compared to urban areas (29%) (Rajan and Sunitha 2015). Among these, rural areas count the most. Kerala had the highest percentage of elderly people living in households, and Dadra and Nagar Haveli had the lowest. In rural areas of Punjab, Goa and Kerala elderly people were living in more than 40% of households (Table 1.4).

Table 1.4 Percentage of households with elderly people by place of residence, 2011

	Total	Rural	Urban
India	31.3	32.5	28.8
Jammu and Kashmir	32.1	31.1	34.9
Himachal Pradesh	35.6	37.2	23.7
Punjab	36.5	40.2	30.7
Chandigarh	20.2	11.7	20.5
Uttarakhand	33.3	36.1	26.8
Haryana	31.8	34.6	27.2
NCT of Delhi	24.5	24.6	24.5
Rajasthan	29.4	30.1	27.1
Uttar Pradesh	33.3	34.7	28.6
Bihar	29.6	29.5	30.3
Sikkim	24.6	27.5	17.0
Arunachal Pradesh	18.8	22.2	9.0
Nagaland	19.9	22.1	14.6
Manipur	26.9	25.5	30.1
Mizoram	23.7	23.6	23.7
Tripura	27.6	27.0	29.0
Meghalaya	20.7	20.9	20.3
Assam	26.7	26.6	27.5
West Bengal	30.4	28.8	33.9
Jharkhand	28.6	29.3	26.3
Odisha	31.8	32.7	27.4
Chhattisgarh	27.0	27.7	24.4
Madhya Pradesh	27.9	27.7	28.3
Gujarat	28.8	31.1	25.9
Daman and Diu	14.7	22.0	12.7
Dadra and Nagar Haveli	14.0	18.3	10.2
Maharashtra	33.4	37.9	28.0
Andhra Pradesh	30.8	33.3	25.5
Karnataka	33.8	38.0	27.6
Goa	36.8	40.4	34.7
Lakshadweep	35.3	33.0	36.0
Kerala	41.1	40.6	41.7
Tamil Nadu	32.1	33.7	30.3
Puducherry	31.3	30.8	31.5
Andaman and Nicobar Islands	21.9	24.6	17.3

Table 1.5 Percentage distribution of elderly people who live alone or in pairs

	Living alone			Living with one elderly partner		
	Total	Male	Female	Total	Male	Female
Total	2.7	6.9	4.8	16.0	13.6	14.8
Rural	2.9	7.5	5.2	16.9	14.3	15.6
Urban	2.1	5.3	3.7	13.9	11.8	12.8

More than half of the total elderly persons live in single elderly households, and another 45% live in pairs. After the death or separation of the spouse, or after the flight of their children on their own wings, some elderly people live alone. In India, 7% of elderly females live alone whereas are only 2.7% of elderly males do so. More elderly females who live alone in rural areas (7.5%) compared to urban areas (5.3%). The state with the most elderly people living alone is Tamil Nadu (14.1%), followed by Andhra Pradesh (13.2%), Chhattisgarh (11.7%), Daman and Diu (10.8%) and Puducherry (8.8%). In these states, there are more elderly females living alone than in other Indian states and union territories. That is, less than 5% of elderly males persons live alone in these states. The percentages of elderly females who live alone in these states are: Chhattisgarh (12.7%), Daman and Diu (13.2%), Andhra Pradesh (14.9%), Tamil Nadu (17.0%) and Puducherry (11.6%); see Table 1.5.

Again, we can state that the elderly people living with a spouse are similar to the elderly people living alone in that they have no access to their children or any other relatives. Thus, we find that about 18.7% (16 + 2.7%) of the elderly population live alone.

Literacy Rate

Among the total number of elderly people, 56.5% were illiterate according to the 2011 Census. Among these, elderly females (71.5%) are in the most backward position compared to elderly males (40.9%). Being illiterate increases the vulnerability of elderly persons. Illiteracy also exposes them more to dangers of financial fraud.

Being susceptible to financial abuse is a general pattern seen in the maltreatment of elderly persons in the world (Giordano and Giordano 1984; Sethi et al. 2011; United Nations 2013). The gap between male and female elderly illiteracy is narrowed as their age increases. Among the literates, most of the elderly people have completed their primary level of education. About 11.6% of elderly females are literate without having gone to school. Due to the nationwide programmes of national literacy mission started in 1988 by the Government of India, some people have become literate without schooling. Kerala achieved 100% literacy through the programmes of this mission. About 14.3% of elderly males have completed their

Table 1.6 Educational level of elderly people in India (%), 2011

		60+	60–69	70–79	80+
Illiterate	Total	56.5	54.3	59.5	61.0
	Males	40.9	38.5	44.3	46.5
	Females	71.5	69.7	74.6	73.8
Literate	Total	43.5	45.7	40.5	39.0
	Males	59.1	61.5	55.7	53.5
	Females	28.5	30.3	25.4	26.2
Literate without educational level	Total	8.7	8.1	9.3	10.4
	Males	7.2	6.8	7.6	8.6
	Females	11.6	10.8	13.1	13.7
Below primary	Total	20.3	18.6	23.2	24.1
	Males	18.3	16.5	21.3	22.8
	Females	24.3	22.8	27.5	26.6
Primary	Total	26.2	26.0	26.7	26.5
	Males	24.8	24.3	25.7	26.2
	Females	29.1	29.5	28.8	27.1
Middle	Total	12.6	13.2	11.4	12.0
	Males	13.3	13.9	12.0	12.5
	Females	11.4	11.9	10.1	11.2
Matric/secondary	Total	12.4	13.2	11.1	10.2
	Males	14.3	15.2	12.9	11.6
	Females	8.7	9.4	7.3	7.6
Higher secondary/intermediate	Total	7.7	8.1	7.0	6.7
	Males	8.6	9.0	7.9	7.4
	Females	6.0	6.4	5.1	5.5
Graduate and above	Total	9.7	10.5	8.6	7.7
	Males	11.4	12.2	10.2	8.9
	Females	6.4	7.0	5.1	5.7

secondary level of education, and another 11.6% have completed graduate-level education. Among the very old elderly (80+), the gap is much less in achieving various educational levels compared to other age groups. For example, at higher secondary level, there was only a 1.9% gap between very old male and female people's levels of attainment. One-fifth of the female elderly in the very old category have attained their matric/secondary and above educational level, while a quarter of the young old females have done the same (see Table 1.6).

Labour Force Participation of Elderly People

Elderly people today live longer and require more care and medical treatment for a greater number of years than in the past. A significant number of elderly people in Kerala who had earlier taken part in mainstream employment are expected to be covered by pension and income security schemes to support themselves in old age. However, this is still not the case in a large number of cases, forcing these elderly to work in the informal sector even into their old age. In India, about 58% of the elderly are out of the labour force. But 31.4% are still working in the mainstream, and another 10.2% are marginal workers. There are comparatively more main and marginal workers among the elderly in rural areas compared to urban areas. Half of the main workers are elderly males. Nagaland has the highest rate of elderly male participation among the main workers (67.3%). This is followed by West Bengal (63.9%), Arunachal Pradesh (62.8%) and Meghalaya (62.1%). Elderly female main workers are also the highest in Nagaland (53.3%), followed by Arunachal Pradesh (36.6%), Meghalaya (31.2%), Manipur (30.3%), Mizoram (29.4%) and Sikkim (21.6%). The greatest proportion of elderly male marginal workers is in Jharkhand (24.5%) followed by Himachal Pradesh (23.5%) and Bihar (21.9%). In Himachal Pradesh (26.5%), Rajasthan (15.3%) and Sikkim (15.8%), elderly females are more engaged as marginal workers than their male counterparts (Table 1.7). Overall, elderly males have continued to participate in the labour force more than elderly females, whereas the latter are more engaged in marginal work than are elderly males.

The figures for non-workers among the elderly suggest that more than half of them may depend on others for their day-to-day care. The situation indicates vulnerability among elderly females (75%), especially those in urban areas. In rural areas most of them are casual labourers. Financial dependency on others increases the vulnerability of elderly persons (WHO/INPEA 2002). Summing up, in India a high incidence of dependency can be observed where the system of social protection is insufficient.

Table 1.7 Percentage of elderly people by work status, 2011

	India	Main workers			Marginal workers			Non-workers		
		P	M	F	P	M	F	P	M	F
Total	60+	31.4	49.4	14.0	10.2	11.0	9.4	58.4	39.6	76.6
	70+	22.1	36.2	8.6	7.4	9.0	5.9	70.5	54.9	85.5
	80+	16.4	27.7	6.4	5.7	7.2	4.5	77.9	65.1	89.1
Rural	60+	34.3	53.0	16.3	12.8	13.5	12.1	52.9	33.6	71.6
	70+	24.3	39.1	10.0	9.2	10.9	7.5	66.5	50.0	82.5
	80+	17.4	28.5	7.3	7.0	8.5	5.6	75.6	62.9	87.2
Urban	60+	24.4	41.0	8.3	4.0	5.1	3.0	71.5	53.9	88.7
	70+	16.6	28.9	5.4	3.1	4.2	2.2	80.2	66.9	92.4
	80+	14.0	25.6	4.7	2.8	3.9	2.0	83.1	70.5	93.3

P means persons

Disabilities Among the Elderly

The proportion of elderly persons with multiple disabilities is higher in rural areas (12.7%) than in urban ones (9.2%). The occurrence of multiple disabilities increases with age and is highest among elderly people in Rajasthan (21%), followed by Jammu and Kashmir (18.3%), Chhattisgarh (16.4%), Sikkim (16.3%), Himachal Pradesh (16%), Assam (15.9%) and Nagaland (15.5%). These states need special attention in the health care sector, especially in geriatric health. It is evident from the table that elderly females are more vulnerable to these disabilities compared to their male counterparts. One explanation for this phenomenon is the extra labour undertaken by females; most female casual labourers are involved in back-breaking work for most of the prime of their lives and after, along with household work.

Prevalence of disability is greater among female elderly persons than males and is explained by the exposure to more risk factors and thus lower recovery rates (Oman et al. 1999). Disability feels awkward and beyond remedy among elderly as they grow older. The 2011 Census 2011 mainly examined four types of disabilities: seeing, hearing, speaking and moving. A quarter of elderly people in India had seeing and moving disabilities, a fifth had hearing disability. Less than 5% had speech problems and about 12% had multiple disabilities (Table 1.8). About 30.5% of the elderly in Karnataka had vision problems (which was 5.2% higher than the national average), followed by Andhra Pradesh (28.4%). Hearing problems among elderly people were highest in Dadra and Nagar Haveli (31.9%).

Dependency, disabilities, loss of spouse, and living alone; concerns related to ageing have not ended, especially among the elderly females in India. According to

Table 1.8 Elderly people (%) by disabilities, 2011

		Total			Rural			Urban		
		60+	70+	80+	60+	70+	80+	60+	70+	80+
Seeing	Total	25.3	25.4	23.7	26.2	26.3	24.5	22.3	22.1	20.8
	Male	23.5	24.1	23.0	24.3	25.0	23.8	20.8	21.0	20.0
	Female	27.1	26.6	24.2	28.1	27.6	25.1	23.8	23.1	21.3
Hearing	Total	19.0	19.5	19.8	18.7	18.9	18.9	20.2	21.6	22.7
	Male	18.7	19.8	20.7	18.4	19.3	19.8	19.5	21.7	24.0
	Female	19.4	19.2	19.0	18.9	18.6	18.1	20.9	21.4	21.8
Speech	Total	3.8	2.9	2.3	3.3	2.4	1.9	5.5	4.4	3.7
	Male	4.2	3.2	2.6	3.6	2.8	2.2	6.0	4.9	4.2
	Female	3.4	2.5	2.1	2.9	2.1	1.7	5.0	3.9	3.3
Movement	Total	25.3	24.9	23.8	25.5	25.0	23.9	24.6	24.6	23.3
	Male	27.6	26.5	24.5	27.9	26.7	24.8	26.6	25.7	23.3
	Female	23.0	23.5	23.1	23.1	23.5	23.1	22.5	23.6	23.2
Multiple disabilities	Total	11.8	15.5	20.2	12.7	16.5	21.5	9.2	12.1	15.9
	Male	10.7	13.9	18.2	11.4	14.8	19.4	8.4	10.8	13.9
	Female	13.0	17.0	21.9	13.9	18.1	23.3	10.1	13.3	17.4

the Census 2011, elderly females are more vulnerable to these issues. It has been observed that women are more likely to be victims of sexual, mental and financial abuse, whereas men are more vulnerable to neglect (Sethi et al. 2011). Though the proportion of elderly people in Kerala is the highest in India, the issues related to ageing are not a risk here compared to other parts of the country.

Kerala Ageing Survey 2013

The Kerala Ageing Survey (KAS) undertaken in 2013 by the Centre for Development Studies (based in Thiruvananthapuram, Kerala) was based on information about the elderly collected in the Kerala Migration Survey of 2011. Of the total 15,000 households throughout Kerala, 5708 were identified as elderly ones in which 7768 elderly people were living. The basic pattern of the elderly in Kerala, according to the survey, was that females outnumbered males. In other words, as in general, with increased age, the share of elderly males shows a decline.

Mental Health

The change in the family structure along with the transformation of finance management in the family results in elderly people losing their relevance and significance in their own house, and increased feelings of loneliness sweeping over them. This has a detrimental influence on their psychological health and can lead to depression. Mental health is also affected by psychological abuse: threat of physical harm, humiliation and generally making the elderly person feel unwanted and demeaned them [Saltzman et al. (2002), cited in United Nations (2013)]. According to the KAS survey, one out of 10 elderly people had depression, and it was high among the very old and females (Table 1.9).

It is interesting to note that the rate of depression was high among the elderly people who have no living children, and the rate was low among those who had two living children. Those who had one living child also had stress in their life. If there were more children, the psychological stress among elderly people was reduced because of the sharing of care, and vice versa.

Table 1.9 Geriatric depression level among the elderly in Kerala, 2013

	60–69	70–79	80+	60+
Low	56.2	48.9	46.2	52.5
Medium	34.9	39.4	38.2	36.8
High	8.9	11.8	15.6	10.7
Total	100.0	100.0	100.0	100.0

Table 1.10 Elderly people and activities of daily living, 2013

	Normal ADL	Need limited assistance	Need complete assistance	Total
60–69	93.0	5.9	1.0	100.0
70–79	87.0	9.7	3.3	100.0
80+	61.3	27.6	11.1	100.0
All	87.1	9.9	3.0	100.0

Table 1.11 Major diseases among elderly people, 2013

	% with diseases
High blood pressure	49.0
Diabetes	27.0
Arthritis	23.7
High cholesterol	16.2
Cataracts	14.0
Cardiovascular illnesses	11.2
Other illness	10.8

Disability

Nowadays in Kerala, elderly people have many roles when they live with their children. Most families include educated and employed women. The grandparents act as care-givers to their grandchildren while the parents are at work. They play a very important role within the household by being support bases for grandchildren, and in turn they receive support from their own children. For this, good health is a necessary precondition which has to be ensured through their active daily living. Most of these care-givers are females.

Table 1.10 describes the extent of elderly people's limitation in physical mobility. About 87.1% of them have no physical immobility. They can carry out their routines without others' help. Only 3% of the elderly are bedridden or physically immobile. They cannot do their daily activities without the help of others. One in 10 very old people need assistance for everything.

Currently the non-communicable diseases such as diabetes, high blood pressure, cardiovascular diseases etc. are common in India, especially so in Kerala. These diseases are also common among elderly people. Around half of the diseased elderly people surveyed in the sample were suffering with high blood pressure (Table 1.11). One in four had diabetes and arthritis. The diseases with disabilities are risk factors to the elderly at their dusk of life.

II

Current socio-demographic data detailed above portray the risk factors and vulnerabilities which elderly people face. In India, as many studies indicate, the abuse and neglect of the elderly (mostly by their relatives and care-givers, people whom they know and trust) affect their quality of life as well as violating their rights to dignity and respect. This emerging tendency is a serious concern. The topic requires deeper and intense academic focus, conceptual clarity, deliberations on the role of professionals in mitigating the problem, and understanding of interventional strategies to combat the issue. This book, the first of its kind, discusses all these parameters and presents a comprehensive work on the topic relevant for a cross-section of the population. We hope it will generate more interest in the field and promote large-scale national study on the issue of abuse and neglect of the elderly.

Of the 14 chapters in this volume, including this introduction, all are from well-known researchers working in the field with many publications on ageing to their credit. Their contributions make this a very special publication as this is the first time the topic of abuse and neglect of elderly people has been dealt with in such breadth and length, covering many different dimensions. Comprehensive academic focus on this subject has so far been given limited coverage in India, mainly because it is still hidden, concealed by the victims, with only a few cases coming to light because of media coverage and a few (generally small-scale with a limited population base) research studies. However, gradually, the issue is being recognized by society at large, with some professionals and administrators in recent years identifying the problem, looking at its incidence and prevalence, and seeking solutions to combat it and minimize its adverse consequences. They are attempting to provide definitional and conceptual clarity on the topic, and as a result more understanding about the victims and perpetrators is emerging. In the last 5 or 6 years, two important surveys on understanding the problem of abuse and neglect of the elderly have appeared in the public domain: one by HelpAge India based on data from various cities in the period 2012–2014; the second by the United Nations Population Fund (UNFPA) as part of its study published as a Report in 2011 on the Status of the Elderly in select states of India.

In recent times, special issues related to abuse and neglect of the elderly, with international and specific national focus, have been published and read in many parts of the world. This timely publication on India puts together current perspectives on the subject for wider readership across sections of the population in this country and abroad. India (with its ageing population being one of the largest, next to China's, in the world) can contribute in many ways to enhancing our understanding of the concept of abuse and neglect of the elderly, and provide useful insights to gerontological discourses. The various chapters provide rich material to further our comprehension of the extent of the problem (who the victims and perpetrators are, among which segments of the population the problem exists, what constitute effective strategies of intervention).

As societies are ageing rapidly and socio-economic changes are taking place (United States Census Bureau 2016), research indicates that elderly people face different forms of mistreatment, including abuse and neglect which encompass any type of behaviour towards them leading to physical aggression, psychological damage, financial or material ill-use, chronic verbal abuse and aggression, neglect and violation of rights (Shankardass 2003). Abuse and neglect of the elderly has been the subject of research in the developed countries for approximately 40 years, with specific country variations, and India and other Asian countries following these have brought the problem to public attention mainly from the turn of the twenty-first century (United Nations 2002; Shankardass 2013).

Changing socio-demographics of India, as outlined above in the first section of this chapter, indicates that with rising numbers of older people in almost all states of India their vulnerabilities and problems are likely to increase. Ageing women and widows are more vulnerable to abuse and neglect and the abusers are mainly caregivers from the family. Migration patterns which leave older people alone in towns and villages also make them vulnerable to abuse and neglect. Further lack of social security and onset of disabilities put the elderly at greater risk. As Mala Kapur Shankardass outlines in Chap. 2, the realities of a rapidly ageing population, along with prevailing ageism, individualism and the breakdown of traditional support systems suggest that older adults will continue to be at risk of abuse and neglect, and potentially at younger ages than in previous generations. To combat the problem, many initiatives by the government and non-governmental organizations are taking a rights-based approach to improving the living conditions of the elderly. In Chap. 3, Khan, theorizing on the issue, indicates the different forms of abuse as seen in the Indian context and further discusses need to conceptualize the problem and understand its intensity. It is time for the country to develop intervention strategies to overcome the problem.

India is generally perceived as a country with a strong tradition of family that nurtures, protects and respects its 'dependent members', including children, women and the elderly. However, in modern times this tradition seems to be weakening with increasing incidence of family violence against all these three types of family member. Anupama Datta (based on inferences drawn from data collected by HelpAge India in various cities across the country in the period 2012–2014) outlines in Chap. 4 the components of elder abuse, its rate of prevalence and most prevalent types. The elderly in the survey express their personal as well as general experiences of abuse and reporting of abuse, and suggest effective measures and sources of information to deal it. In Chap. 5, Varughese and Jamuna in presenting the anatomy of elder abuse describe its occurrence at all economic levels and among families, communities and institutions. Abuse and neglect may be intentional or unintentional, and assessment of abuse becomes difficult if the victim has cognitive impairments and is unable to describe the experience. The authors put the onus for reporting abuse and neglect in institutions and the community on the care-providers, who must be proficient at assessing the situation. Through case studies the authors discuss the epidemiology off elder abuse and strategies for assessment, evaluation and early intervention.

Irudaya Rajan and Arya in Chap. 6 discuss forms of abuse and neglect of the elderly by families based on the KAS. They examine the risk factors for it happening within the family, and explain the relationship of the abuser to the abused elderly person. As they explain, there are not only visible signs of abuse and neglect but many (generally lifelong) invisible ones in the form of pain, mental torture and distress. In Chap. 7, by S. Siva Raju and Nidhi Gupta, the authors reflect on the problem in Maharashtra and attribute its rising incidence to the changing social, familial and economic context caused by demographic transition in India. They explore the phenomenon of elder abuse, its forms and extent, the role of perpetrators, and policy and programmatic implications for minimizing it.

Chapter 8 by Muthuvenkatachalam Srinivasan and Sandhya Gupta on the prevalence of elder mistreatment in Uttarakhand state in Northern India depicts neglect and exploitation as the most common type of mistreatment, and finds these are positively associated with psychological distress and being aged over 80. They are negatively associated with social support availability. Kumari and Sekher in Chap. 9 highlight the case of widows in the rural areas of Jharkhand state, who are seen as a burden on families and become vulnerable to abuse and neglect. Illustrating their chapter by empirical evidence, the authors reveal the extent of deprivation and isolation faced by the widows, showing various dimensions of abuse and neglect. Sebastian and Sekher in Chap. 10 make use of data from the UNFPA survey done in 2011 in seven states of India which have comparatively large proportions of older people. These data reveal a disturbing situation in India where about 10% of older people indicate they are victims of abuse and neglect, with verbal abuse and disrespect being the most frequent forms and the main perpetrators being sons (41%) and daughters-in-law (32%). The authors raise doubts about families in particular, and Indian society in general, as safe places for older people.

Chapter 11 by Selim Jahangir, Nikhil P. N., Ajay Bailey and Anindita Datta, while focusing on abuse and neglect of the elderly in homes, looks also at institutions, a growing concern especially in urban areas. Based on their qualitative studies conducted in Delhi, Kolkatta and Kerala, they highlight different forms of abuse which have affected the living arrangements and everyday care of older men and women. Goel, Koolwal and Jain in Chap. 12 discuss yet another aspect of elder abuse and neglect, the growing incidence of it among HIV-positive people. As outlined by the authors, numerous such cases remain unspoken of and unnoticed, perhaps because victims tend to refrain from complaining or resist help due to their dependence.

Asha Banu Soletti and Pankhuri Bhatia in Chap. 13 take the reader in another important direction; that of interventions, which are essential to combat the problem. Given the paucity of care services for elderly people in India, it is imperative that existing professionals (such as counsellors, doctors and social workers) who interface with the elderly in different contexts be trained to handle cases of abuse and neglect. Outlining the role that social care and health professionals currently play to mitigate elder abuse, the authors argue for investment in building the capacity of existing professionals to identify, screen and refer cases in order to improve intervention. Certain recommendations for strategies to build mechanisms for creating conducive spaces for the elderly within the existing system are suggested. In Chap. 14 Kuba emphasizes prevention and management of elder abuse and neglect in society. Elucidating a few strategies to be adopted, she sees similarities in them across different parts of the world as principles of prevention and management founded on a rights-based approach. Strong policy and an appropriate framework for implementation of laws and programmes to prevent and manage abuse and neglect of elderly people are necessary in societies. Training of professionals to deal with such cases is required immediately, along with creating awareness of the problem in society. Research is another component required to understand the problem in its multi-dimensional perspectives.

III

The first section of this chapter examined the demographic, socio-economic and health patterns of elderly people in India and Kerala by using the respective data from 2011 Census and the Kerala Ageing Survey 2013. The second section focused on elder abuse and neglect among different segments of the population, and both sections together have indicated the seriousness of the problem, incidence of which is growing in the face of few mitigating interventions and policy responses. From the socio-demographic picture given above we can generalize the main issues related to the elderly population. The basic problem in Indian society is the lack of security at three major levels (financial, well-being and psychological) with lengthened years of life and particularly life-style diseases amplifying morbidity along with the cost of living. Some pertinent questions arise as to what responsibility the country has for taking care of these elderly people. How do we equip families to care for the elderly from a rights-based and humane approach? What kind of training should be given to social and health-care professionals to minimize the risk of abuse and neglect of the elderly? What kind of policies and programmes should the government develop to make the later years of men and women more meaningful, more productive in all aspects and provide a better quality of life? How do we achieve a sound system which offers multiple solutions to the problems of old age?

In India, universal coverage of social security for the elderly may not be practical. So it is based on a targeted approach such that old people fall under no income or low-income group. The targeted group of elderly are benefited from the scheme, though it is inadequate for their daily living. But the elderly are not a homogenous group. They include young old people (aged 60–69), old old people (70–79), very old people (80 and over), widows, disabled and bedridden people etc. All need different kinds of help and support according to their requirements. One of the basic solutions to the problem of elder care, including minimizing the risk of abuse and neglect, is to make older people emotionally and financially secure, protected and healthy. Government, civil-society members and professionals with an interest in the welfare of older people need to join together to improve their well-being across the country.

References

Giordano, N.H., and J.A. Giordano. 1984. Elder abuse: A review of the literature. *Social Work* 29 (3): 232–236. JSTOR (23714248).

Irudaya Rajan, S. and S. Sunitha. 2015. Demography of ageing in India 2011–2101. *Helpage India Research and Development Journal*. 21 (2): 13–22.

Oman, D., D. Reed, and A. Ferrara. 1999. Do elderly women have more physical disability than men do? *American Journal of Epidemiology* 150 (8): 834–842. doi:10.1093/a010088.

Sethi, D., et al. 2011. *European report on preventing elder maltreatment*. Denmark: World Health Organization. Regional Office for Europe.

Shankardass, Mala Kapur. 2003. Combating elder abuse in India: An emerging social, legal and public health concern. Symposists' Papers, Invited Symposia on Elder Abuse, Executive Editor, Toshio Tatara, Tokyo International Forum, Japan.

Shankardass, Mala Kapur. 2013. Addressing elder abuse: Review of societal responses in India and selected Asian countries. International Psychogeriatrics, 1–6, International Psychogeriatric Association.

United Nations. 2002. Department of Public Information, DPI/2264, Report of the Secretary-General Kofi Annan released at the Second World Assembly on Ageing, Madrid, Spain 8–12 April 2002.

United Nations. 2013. *Neglect, abuse and violence against older women*, 1–48. New York: United Nations Publication. Available from http://undesadspd.org/Ageing.aspx.

United States Census Bureau. 2016. An Aging World: 2015, International Population Reports, U. S. Department of Commerce, Economics and Statistics Administration.

WHO. 2002. *Missing voices: Views of older persons on elder abuse*. Geneva: World Health Organization.

Chapter 2
Perspectives on Abuse and Neglect of the Elderly in India

Mala Kapur Shankardass

Contextualizing the Issue

Elder abuse and neglect as social, legal and public health problems have now been recognized all over the world for more than four decades, though it is only at the turn of the twenty-first century that they emerged as a serious, pervasive problem to confront and combat. Today, nearly all international conferences, academic meetings and civil society members involved with ageing issues and gerontological discourses are holding discussions and symposia on the theme, and much new research providing deeper understanding of the problem and interventional strategies are being shared at these forums. Many nations are developing tools for detection and screening, and are building resources to prevent and manage the problems. The paucity of data on the topic which was a feature of the twentieth century is being overcome. The *Journal of Elder Abuse and Neglect*, published in the USA, is becoming popular all over the world, and new scientific journals (like the UK-based *Journal of Adult Protection*, which emphasizes evidence-based practice in relation to safeguarding adults) are increasing their readership across countries. In India, too, elder abuse and neglect have caught the attention of researchers, social workers, academicians and policymakers, and in recent years there have been responses from various quarters attempting to understand the problems, prevent and manage them. Since the adoption of the United Nations' (UN) *Madrid International Plan of Action on Ageing* by the World Assembly in 2002, many UN member countries have shown a growing interest in policy response and interventions to elder abuse and neglect, even though incidence and prevalence of elder abuse varies in them. Since then, many more studies in different countries, including India, have been initiated to understand the problems and

Mala Kapur Shankardass (✉)
Department of Sociology, Maitreyi College, South Campus,
University of Delhi, New Delhi, India
e-mail: mkshankardass@gmail.com

Mala Kapur Shankardass and S. Irudaya Rajan (eds.), *Abuse and Neglect of the Elderly in India*, https://doi.org/10.1007/978-981-10-6116-5_2

capture their various dimensions as the number of older people increases. It is assumed that with population ageing, occurrence of elder abuse and neglect will also rise.

In 2002, the World Health Organisation (WHO) published the Toronto Declaration, which offers the most consensual definition of elder abuse, and it has been used in a number of countries since then:

> a single, or repeated act, or lack of appropriate action, occurring within any relationship where there is an expectation of trust which causes harm or distress to an older person. (WHO 2002)

However, this definition is still discussed and contested, and a few countries have come out with their own understandings of elder abuse and neglect. For instance in Quebec in Canada, the term 'elder abuse and neglect' has been replaced with 'mistreatment of older adults' and it includes violence and neglect. In the USA, too, 'elder mistreatment' as a term is often referred to in cases of elder abuse and neglect. Many studies conducted in the Indian context, typically classify elder abuse in five broad categories, and refer to actions against elders perpetrated by someone who is trusted. These categories include physical abuse, emotional or psychological abuse, financial abuse or exploitation, neglect, and sexual abuse. In many societies, especially those with a substantial ageing population, a category recognized as self-neglect (where elders are unable to provide for their own health and safety and for which no perpetrator has been identified) is also being recognized, but in India this remains an unrecognized field. A universal definition of elder abuse and neglect acceptable to a cross-section of the population is also missing and besides the above-mentioned five categories, abandonment, isolation, intimidation, fiduciary abuse, extortion, unreasonable confinement, active versus passive neglect and coercion are also identified as forms of elder abuse. More recently, intimate partner violence (i.e. domestic abuse of older women) is catching the attention of professionals and the lay public (Winterstein and Eisikovits 2014). Lack of mandatory reporting laws, which require certain individuals and professionals to report instances of abuse to an official agency or to the police also add to the lack of clarity on an acceptable, universal definition of elder abuse and neglect in the country.

In India, there has been a limited political response to this social problem even though the 1999 policy on older persons and the revised 2011 draft policy (which has still to be approved by the Cabinet) mention the concern (MoSJ&E 2011). The Maintenance and Welfare of Parents and Senior Citizens Act, 2007 takes care of the neglect of older persons in the country. The Act has been adopted by many states and is also being implemented in some of them, but not all states have yet taken the five steps required for its implementation; namely, they must notify the Act, frame rules for the Act, appoint maintenance tribunals, appoint maintenance officers, and appoint appellate tribunals (Shankardass 2012). Further, very few older people use this legislation to seek justice for themselves and when they do, the legislation has provisions only to provide maintenance, financial help and social security. It does not pay adequate attention to social care aspects. It overlooks the need for emotional support and love as well as at times physical care from the family. The Act also ignores the abuse and neglect of senior citizens in other than family circumstances. This Adult Protection Legislation in India, as in other countries (namely Britain, Canada and the USA) has been critically reviewed, being seen more as an ageist response to a complex problem (Shankardass 2007).

Across the country, for the last few years various non-governmental organizations (NGOs) and professional groups consisting of researchers, academicians, practitioners and advocates have come up with ways to prevent elder abuse. As in many other countries, these organizations work as small grass-roots initiatives concerned with occurrence of elder abuse in the community and finding ways to protect the safety, security and dignity of older persons. They try to provide support to victims of elder abuse but are often hindered by withdrawal of complaints by older people themselves. However, there is no major group or organization which is committed solely to handling elder abuse cases. The International Network for Prevention of Elder Abuse (INPEA) with its regional and national representatives as an advocacy organization has raised awareness on the issue in the country as well as 64 other countries. The INPEA India chapter, active since 2002, has been raising awareness among professional groups, service providers, and the lay public on the need to recognize elder abuse in society, to make it visible, and to support victims of abuse and neglect. INPEA in 2006 declared 15 June as the date on which to observe World Elder Abuse Awareness Day (WEAAD) every year. Since then, the INPEA India chapter has focused on the prevention, detection and management of elder abuse and neglect. Various programmes to prevent and manage elder abuse and neglect are organized by different groups in many parts of the country to support the INPEA initiative. For the last few years the government, under the aegis of the Ministry of Social Justice and Empowerment (the nodal agency concerned with care of older people), has used 15 June to take out advertisements in the national newspapers asserting the need for dignity and social security of older people. In 2011 the trend was set for the first time by the Delhi Government when it took out a half-page advertisement in newspapers clearly stating the announcement of the World Elder Abuse Awareness Day and its commitment to the health and well-being of all senior citizens, and promising to continue improving and expanding its services to make older people's golden years happier and dignified. Significantly, the government has announced the neglect and abandonment of senior citizens as a cognizable offence punishable with a fine of Indian Rupees 5000 or 3 month's imprisonment or both. The government now has a slogan 'A society that cares for its elderly, is a caring and loving society'. Since the inception of WEAAD, police, NGOs and other civil society members have participated in marking their concern for prevention and management of elder abuse.

Across the world it is now recognized that while efforts to address elder abuse and neglect are increasing and more resources are being invested as a societal response, the realities of rapidly growing older population, along with prevailing attendant ageism, individualism and breakdown of traditional support systems, suggests that older adults will continue to be at risk of abuse and neglect, and potentially at younger ages than in previous generations (Lachs 2010). With feminization of ageing taking place in most countries, older women (living longer and more numerous) are more vulnerable to abuse. Especially ageing widows and frail as well as disabled older people are at greater risk of abuse and neglect, not only by family members but also by non-family members and unknown people. Initially, research on elder abuse and neglect by family members indicated care-giver stress as the cause for the problems, but later studies revealed many other characteristics of perpetrators, such as mental health and behavioural problems, drug abuse, family disputes, intimate partner abuse, etc. With regard to non-family abusers, untrained and unscrupulous care-givers, financial tricksters and exploiters, and petty criminals are recognized as people of whom older people have to be careful. At the community level, prevailing ageism is seen to devalue and exploit older people leading to their abuse in various situations and circumstances.

In India, as in other countries greater attention is being paid to abuse and neglect of older people residing in the community, either living with others or alone, and to those residing in institutions such as old-age homes or assisted-living facilities. However, there is still no emphasis on measurement scales and indexes for elder-abuse instances and episodes. Although some attention to this is being paid in some developed countries, internationally accepted, standardized measurements require further development. While families are held responsible for mistreatment and neglect, there is still hesitation in suing institutions such as old-age homes for non-compliance of their liability and obligation to take care of their older residents. However, through the Indian Contract Act 1872 and the Law of 'Torts', any mistreatment and failure in caring can result in a civil suit being filed against the management of the institution for breach in contract to care, even in the absence of a written contract, and damages can be claimed for the wrong done (Bakshi 2000a). There is a need to take seriously the provision in the Indian Penal Code under which breach of contract may attract criminal liability against old-age homes and senior-citizens housing complexes, which are fast growing in all the towns and cities of the country. However, institutions for meeting the needs of older persons have not so far been brought under legal purview in order to alleviate elder abuse.

Since the independence of the country, government care of old parents by children has been a crucial issue. For a long time, The Hindu Adoption and Maintenance Act, 1956 secured the rights of care for the parents by the children, being applicable to both, sons and daughters, married or unmarried. The law relating to Muslims and other religious groups is not codified in India but the position regarding them is substantially the same. It is contained in the Code of Criminal Procedure, 1973 (even though the provision is really of a civil nature) in Section 125 (1) (2) which recognizes the right of parents without any means to be supported by their children having sufficient means (Bakshi 2000b). Thus, there is a comprehensive statutory provision as to maintenance which is not confined to persons of any particular community. However, 'maintenance' under Section 125 of the Code of Criminal Procedure, 1973 has not been defined, which gives leeway to the 'fair' completion of the case and suitable grant of maintenance for maintaining an adequate and appropriate standard of living as per the status of the old parent. While in settlement of some disputes liberal interpretation of maintenance has been adopted in the courts (Bakshi 2000b), it not being confined to food, clothing and shelter only, but including expenses for medical attendance also, the questions of justice and equity remain disputed even though the provision is intended to exclude destitution and vagrancy. In addition there are the issues of complex, delayed and lengthy procedures which get in the way of giving relief and combating abuse (Shankardass 2010). In the recent past, in cases of old parents being deserted by children or being ill treated, the Maintenance and Welfare of Parents and Senior Citizens Act, 2007 has been used as a legislative measure.

There is a need for more attention to adult protection legislation, along with advances in the creation of protocols for detection, intervention and programming. It is observed that while on the one hand awareness about legislations in place to protect the interests of older persons is low, on the other hand there are difficulties being experienced in using legal measures for relief from abuse. There is hesitation on the part of older persons to sue their family members and institutional managers for breach of contract in caring. Empowering the older persons to demand and have their needs fulfilled within the human rights framework is a challenge not only in India but also in the rest of the world. Elder abuse and neglect needs more visibility and media attention and awareness among the families and communities, and only then can they be stopped, prevented and managed.

Research Orientations

In India, there has been no national prevalence study so far to assess the magnitude of elder abuse and neglect, but there have been small-scale studies which indicate certain aspects of the problems but not the frequency and types of abuse, nor the characteristics of perpetrators. Systematic data on this would help to identify gaps in the detection of abuse and the need for various types of support systems. More precisely, the need for services, and the types of services that would be most helpful, could be ascertained in part by determining both the *incidence* of abuse—how many new cases are there over a period of time, such as a year—as well as the *prevalence* of abuse, including not only new cases but those that are still ongoing and have not been resolved. It would be helpful, but unrealistic because of the cost, to have such information available not only at the national level but also for specific states and localities. However, prevalence studies are actually crucial to solving the problem of elder abuse. A rigorous and representative prevalence study has huge implications for policy, practice, the law and further research.

UNFPA, India—as part of its programmatic exercise on ‘Building knowledge base on population ageing in India (BKPAI) in 2011’, in collaboration with the Institute of Social and Economic Change (ISEC), Bangalore, the Institute of Economic Growth, Delhi and the Tata Institute of Social Sciences (TISS), Mumbai—conducted a seven-state study which also documents the incidence of elder abuse and neglect by family and non-family members. This provided some of the first population-based data from rural and urban areas to throw light on the various dimensions of the problem with state comparisons. Some 9852 older people aged 60 years and above were interviewed from 8960 households across states which have comparatively high proportions of older people, namely Himachal Pradesh, Kerala, Maharashtra, Odisha, Punjab, Tamil Nadu and West Bengal (UNFPA 2012).

Besides this, HelpAge India, a national-level NGO, has conducted yearly studies since 2013 which also reveal the different aspects of abuse and neglect of older people in the different states of the country. While through the various studies we are able to recognize the perpetrators of abuse and neglect and also to some extent the impact on the victims, and certain interventional strategies, there is no emphasis on standardized intervention approaches and on tools which can support these tasks. If there are adequate protocols for prevention, detection and management of elder abuse and neglect, as there are in some developed countries, it would help various stakeholders in addressing the problem. For instance, issues related to elder abuse, including research, prevention, detection and treatment are addressed in the USA at the state level, by several federal agencies, by non-profit organizations and various ad hoc coalitions dealing with specific topics, and in over 3000 counties across the 50 states (Thomas 2018, forthcoming). But, such a response in India has yet to be geared up. Not enough has been done to prevent elder abuse and neglect and to establish appropriate response services to manage the problem. It is hoped that with the forthcoming. full. national-scale. longitudinal ageing study in India (LASI)—being coordinated by the International Institute for Population Sciences in collaboration with Harvard School of Public Health and the University of Southern California, USA—aspects of elder abuse and neglect will also be scientifically highlighted with sound methodological rigor while studying health, economic, social dimensions and determinants and consequences of population ageing in the country.

Missing Links

In India there is no national strategy for the response to and prevention of elder abuse and neglect. Nor are the problems addressed from a practice-delivery stance. There are no increased efforts to improve education among professionals and service providers about what constitutes elder abuse, how to act on it and how to support at-risk older people. We need to focus attention on proactive outreach services to meet the needs of vulnerable older people—no one agency currently has a coordination role. We also need improved community awareness for older people about their rights and legal options in situations of harm or abuse to ensure they live free from violence. In general, developing a comprehensive service system to deal with victims of abuse and neglect and strengthening public policy for combating elder abuse and neglect are the needs of the hour. Pillemer et al. (2015) have identified another important challenge for improving research knowledge to prevent elder abuse and neglect and to support victims. This requires an interdisciplinary response, the importance of which has yet to gain recognition in the country. Developing sound research methodology for identifying and screening cases of abuse and neglect by different members of the family and by non-family members is a foremost requirement. Further, theorizing the concepts requires developing theoretical orientations for conceptual clarity. In achieving a comprehensive theoretical framework, many difficulties arise from the use of varying definitions of elder abuse, variations in methodologies used to study elder abuse and neglect, and different measurement instruments.

Much research into elder abuse and neglect in India deals with the family violence model with little attention being paid to abuse in residential and institutional settings. This is a growing area of interest as the number of old-age homes, care homes and residential complexes for older people is fast increasing in the country, with no proper guidelines for the management and organization of these places. Research on elder abuse and neglect in residential settings in developed countries (namely the USA, Britain and Canada, including stay provisions in hospitals, nursing homes, residential care homes and long-stay homes) indicates that since the later part of the last century factors of quality care, health, hygiene and nutrition along with concerns related to over- and under-medication are gaining relevance (Glendenning and Kingston 1999). In India this is an emerging field requiring theoretical and empirical understanding and research tools to assess the situation. There is a need to move away from anecdotes to scientific evidence to understand resident and patient abuse of older people in institutional settings, which seems to be becoming widespread in different parts of the world including India. Increasing incidents of physical and psychological abuse, mistreatment, assault and neglect in residential settings merits concern and requires working at the policy level for an 'abuse-free environment'. We need deeper, intensive and extensive studies of elder abuse and neglect in residential as well as in domestic settings.

In many Indian states police have played an active role in preventing elder abuse and neglect, but not much has been documented. Police have intervened with families to protect and safeguard older people's right to life with dignity. Police also make regular visits to older persons' homes, especially those living alone, and interact with them in order to keep an eye on their interactions and activities. Police's contribution as an intervening agency to prevent elder abuse and neglect has been recognized in a few countries. For example, in the Canadian province of Montreal especially, police, in synergy with their partners, are required to play a crucial role in countering mistreatment of older adults (Beaulieu et al. 2018). This requires a national approach with increased collaboration between service providers, research organizations, legal services, health-care providers, and anyone involved with the care and well-being of older people within our communities.

Concluding Comments

Elder abuse all over the world is a multi-layered, multi-dimensional concept still requiring clarity of definition, tools for detection and measurement, adequate diverse interventional approaches which protect and support people in their later years. Variations in gender, economic position and physical condition, which make different older people vulnerable and at risk of elder abuse and neglect are now recognized (Berkman et al. 2012) but much more needs to be done to prevent and combat it. There is sufficient research information to demonstrate that the study of elder abuse and neglect needs to embrace not only the family violence model, and strategies to stop it in domestic settings, but also the paramount need to better train care-givers, reduce their stress levels, and improve home and institutional conditions. This requires an inter-disciplinary response, cross-national understanding of social, legal and public-health aspects of the problems from a human-rights perspective. Older people have a right to a life of dignity, respect and esteem and this in no way can be denied to them. Rightly, the Human Rights Commission has identified that older persons have a right to have their human rights protected. Governments and civil-society agencies at both micro and macro levels have to and must develop services and have provisions to combat elder abuse and neglect.

References

Beaulieu, Marie, Michelle Côté, Joséphine Loock, Monia D'Amours, Luisa Diaz, and Jacques Cloutier. 2018, forthcoming. The Montreal police model: Actions to counter mistreatment of older adults. In *International handbook of elder abuse and mistreatment*, ed. Mala Kapur Shankardass. Berlin: Springer.

Bakshi, P.M. 2000a. Homes for the aged—legal aspects. In *Lecture series in geriatrics, health care promotion trust*, ed. P.C. Bhatla, 416–417. New Delhi: National Institute of Primary Health Care.

Bakshi, P.M. 2000b. Senior citizens and the law. In *Lecture series in geriatrics, health care promotion trust*, ed. P.C. Bhatla, 415. New Delhi: National Institute of Primary Health Care.

Berkman, L., T.V. Sekher, B. Capistrant, and Y. Zheng. 2012. Social networks, family, and care giving among older adults in India. In *Ageing in Asia: Findings from new and emerging data initiatives*, ed. J.P. Smith, and M. Majmumdar. Washington, D.C.: National Academies Press.

Glendenning, Frank, and Paul Kingston (eds.). 1999. *Elder abuse and neglect in residential settings: Different national backgrounds and similar responses*. Binghamton: The Haworth Maltreatment & Trauma Press.

Lachs, M. 2010. *Treat me, not my age: A doctor's guide to getting the best care as you or our loved one gets older*. New York: Penguin.

MoSJ& E. 2011. Ministry of Social Justice and Empowerment, Government of India. http://socialjustice.nic.in/pdf/dnpsc.pdf.

Pillemer, K., M.T. Connolly, R. Breckman, N. Spreng, and M.S. Lachs. 2015. Elder mistreatment: Priorities for consideration by the White House Conference on Aging. *The Gerontologist* 55 (2): 320–327.

Shankardass, Mala Kapur. 2007. Maintenance and Welfare Bill, 2007 as adult protection legislation: A critical assessment'. *Research and Development Journal* 13 (3), HelpAge India.

Shankardass, Mala Kapur. 2010. Elder abuse and adult protection. In *Ageing and health in India*, ed. C. Shanthi Johnson, and S. Irudaya Rajan. Jaipur: Rawat Publications.

Shankardass, Mala Kapur. 2012. Combating elder abuse: Need for comprehensive right based action. *The Tribune*, 15 June.

Thomas, Cynthia. 2018, forthcoming. Coordinating elder abuse prevention and treatment across organizations in the United States. In *International handbook of elder abuse and mistreatment*, ed. Mala Kapur Shankardass. Berlin: Springer.

UNFPA. 2012. *Report on the status of elderly in selected states of India-2011, Building a knowledge base on population ageing in India*. New Delhi: UNFPA.

WHO. 2002. *The Toronto declaration on the global prevention of elder abuse*. Geneva: World Health Organisation.

Winterstein, Tova Band, and Zvi Eisikovits. 2014. *Intimate violence across the lifespan: Interpersonal, familial, and cross-generational perspectives*. Berlin: Springer.

Chapter 3
Theorizing Elder Abuse in the Indian Context

A.M. Khan and Nasir Ahmad Bhat

The greatest challenge of the twenty-first century is an increasing number of older persons and their social security, emotional health, psychological, economic supports and general care to safeguard their basic needs and uphold the dignity of human life. It would be shameful to any family, society and nation if people blessed with longevity were bedridden and left to breathe without quality care. It would be a black spot on the glory of economic and technological breakthroughs if millions of families were left to struggle with the miseries of longevity and fail to cope with daily essential activities of life. The real challenge is to provide space and basic facilities to millions of people living lives marred by abuses of various kinds. The trivial challenge is to provide adequate social, emotional, economic and health-care support to the elderly. The much greater challenge concerns older women who are culturally care-centric regardless of their own health status. They need enormous resources for general, medical and nursing care. The most difficult challenge is how to retain and restore the 'traditional culture of elderly care' when market forces are pushing self-centric, pleasure-seeking tendencies amongst younger generations and slowly reducing their concern for people other than themselves. In the absence of meeting such challenges for the growing population of older people, the occurrence of elder abuse is an inevitable reality; we simply must use scientific tools to acquire data in order for us to assess the quantum of these problems (Khan 2004).

Elder abuse dates back to ancient times, but it came to notice only after the concepts of child abuse and family violence were developed in the USA during the last quarter of the twentieth century. Before that 'it remained a private matter, hidden from public view; it was seen as a social welfare issue and subsequently a problem of age difference' Burston (1975). Nowadays, elder abuse is recognized as

---

A.M. Khan
Center for Jawaharlal Nehru Studies Center, New Delhi, India

N.A. Bhat (✉)
Department of Psychology, Jamia Millia Islamia, New Delhi 110025, India
e-mail: bhatnasir86@gmail.com

Mala Kapur Shankardass and S. Irudaya Rajan (eds.), *Abuse and Neglect of the Elderly in India*, https://doi.org/10.1007/978-981-10-6116-5_3

constituting public health and criminal justice issues in some of the developed nations. In recent years, worldwide concern about human rights, gender equality, domestic violence and population ageing is evident everywhere, in both developed and developing nations. The research is being undertaken globally on abuse. However there are many grey areas for exploration and innovations. So, cross-cultural research studies provide an impression that the nature and types of elder abuse are specific to country and culture. Some forms of abuse are found across the countries. For example, neglect, emotional abuse, and mistreatment are mentioned in many studies. Elder abuse was first described in the *British Scientific Journal* using the term 'granny battering' (Baker 1975; Burston 1975), during the 1980s. Some scientific research and government actions in Australia, Canada, China, Norway, Sweden and the USA (and later in Brazil, Chile, India, Israel, Japan, South Africa, the UK and European countries) has been initiated.

'Abuse' is a generic term which is used as a verb as well as a noun. It means improper use of something. It also includes regular or repeated cruelty in treatment. Different terms are used for abuse. These are: misuse, ill-treatment/maltreatment, and mass misuse of anything: country, person, places both living and non-living like country abuse, animal abuse, law abuse, office abuse, or generation abuse, etc. When ill-treatment/maltreatment is given to an older person, it is termed abuse. It represents 'all types of mistreatments, abuse and behavior towards older adults' (Wolf 1996). Wolf defined elder abuse as acts of commission (intentional behaviour) and omission (failure to act), Elder abuse is harm to older adults. In the literature, several forms of abuse are found: physical, medical, psychological, sexual, abandonment, neglect (material, self-neglect), violation of rights.

In every society there are social and cultural beliefs and practices regarding the roles of family members towards each other. The life of each member in the family is embedded into different types of concern, care practices and expectations of each other, which over the years have been socially and culturally transmitted in the minds of people in the society with socio-political transitions and development of human rights. Under the influence of industrialization, urbanization and modernization, the nucleus of relationships has changed rapidly. Community-centric society has gradually been replaced, partially and fully, by individualistic society. The dynamics of relationships in the family determine the various profiles of elder abuse. Different terms evolved defining elder abuse. Chronological development of terminology can be seen in: granny battering, elder abuse, elder mistreatment, battered elder syndrome, elder maltreatment, granny beating, inadequate care of elderly, granny abuse. Whether behaviour is termed abusive, neglectful, or exploitative probably depends on how frequently the mistreatment occurs, how long and how severe it is determine the nature of abuse. For example, in the Navajo community in the USA it is regarded as a cultural duty to share material belongings among the family, something which a researcher conceptualized as economic exploitation. Native American tribes in the USA view elder abuse as a community problem rather than an individual one.

A review of the literature reveals considerable cross-cultural differences in the definition of abuse. The common point is any verbal or non-verbal gesture which harms old adults. Ideas or ideologies which hurt the elderly and provide suffering are considered elder abuse. One of the most commonly accepted conceptual definitions of elder abuse was proposed by Johnson (1995): 'It is state of self or other inflicted suffering unnecessary to the maintenance of the quality of life of elder persons'. The suffering is defined as intense, sustained pain and anguish. This definition focuses on whether the older person has experienced pain and suffering to determine? Whether mistreatment has occurred? The culture differences are built into it because what is normal behaviour in one culture is considered abuse in another. For example: in the Native American set-up, in slum areas where use of abusive language is common, people living there may or may not consider it elder abuse, unless suffering is inflicted. But use of the same language in a higher-class area may be considered a serious form of elder abuse. The Council of Europe Report recommends that consideration should be given to the fact that the differences between acceptable and abusive actions are questions of conscience. The definition depends on the subjective value judgment of researchers, the complexity in conceptualizing, elder abuse multipurpose definitions were proposed by Aber and Zingler (1981).

Conceptualizing Elder Abuse

Elder abuse looks is a social and cultural phenomenon. Human behaviours are shaped differently in different societies. In an individualistic society where people lead independent lives, the elderly may not report abusive behaviour unless they suffer physically. The concepts of loneliness and marginalization in such a society may not be the same as those in developing countries where elderly care is recognized as a social and cultural ethos. And if children don't pay attention, the elderly person may become emotional and take it to heart, gradually becoming negative about their own ability to manage the problem. If for example, an old person develops a cataract, and if children fail to provide treatment from the recognized places/persons, it is taken as ignorance and indifference on the part of the children and older adults react to it. Similarly, social interaction is an equally important need. If elders are deprived of it, they develop emotional problems. In Indian culture, it is essential to make elderly people feel that they are part of the mainstream of family decisions; they are part of all activities and ceremonies in the family. Respect for them by everybody in the family is one of the prominent expectations amongst elderly people. The following statements are illustrations intended to work towards a comprehensive understanding of elder abuse.

- If nagging, commenting, condemning are done to elder adults, it is abuse.
- If elderly people are not served a meal in time, if meals are not according to their choice and interest, and are served erratically, it may be abuse.
- If elderly people are forced to live outside the house, if they are rebuked, compelled to change their habits and way of living, it may be abuse.

- If economic security and health-care are not provided despite repeated requests, it may be abuse.
- If persons are suffering with Alzheimer's, memory loss, forgetfulness and they are made mockery of, it may be severe abuse.
- If elderly people are compelled to carry out activities and hold responsibilities of their own, and are put into a strict discipline of living, eating, dressing, sleeping, walking and talking which they don't like, it is abuse.
- If they are consciously blackmailed, or cajoled for ulterior motives, it may be abuse.
- In the hospital, if the services are not designed to maintain the functional physical status of elderly, and if old persons are not assisted in making frequent, necessary trips to hospital and medical care providers, it is abuse.

Abuse can occur within the family, in an old-age home, in a geriatric hospital or in terminal care institutions like hospices, Elder abuse can be of different types—physical, social, emotional and economic. Neglect/abandonment is also a major form of abuse, though some agencies, including the United Nations would like to consider this as separate because it needs to be considered ion its own merits. Perhaps the most violent and extreme form of elder abuse comprises atrocities/criminal assaults. All the forms of elder abuse are on the increase and therefore should be a matter of concern to the academicians, administrators and policymakers. The problem is very much pertinent in the present world where life expectancy and the proportion of elderly people in the population are both increasing (Nayar 2004). Abuse, neglect and violence against elderly people are quite common. They often go unreported and there are not relevant statistical figures to bring out the nature and extent of abuse. Even where human-rights reports are becoming more explicit, senior citizens do not feel safe and secure as a result of increasing and varied (social, psychological and financial) crimes against them.

The changing value and norms of society are making the problem worse, and the police and law find it very difficult to combat the abuse of elderly people (Gokhale 2006). Elder abuse is a social problem and care-giving should become a social action programme. Gokhale gave several accounts of older men and women who made adjustment in the last phase of their lives. Bhoite (2004) examined how the elderly are marginalized into a category that is unwanted. Society makes them succumb to a state of meaninglessness and emptiness. Family relationships become strained. Chakravarthy (2004) opined that elder abuse occurs when older people have nothing to offer. They possess coping mechanisms, but when the stress increases they cannot adjust to the fast-changing situation and thus abuse becomes covert. Vadakumcherry (2004) noticed an inevitable decline in life dignity, honour and respect during old age. He identified several areas in which the elderly were abused. Some of them as members of a nuclear family consider old age as a burden; and they seek to avoid responsibility; nobody wants to talk to the elderly and consequently they feel very lonely; they do not get the food they like, are not consulted in decision making, are ridiculed, mocked and made a laughing stock or considered a nuisance. Elderly people were not greeted with respect. There were restrictions in the home, and when one took responsibility for something it was permitted as a charity which others lamented.

The atrocities against elderly people are due to economic reasons; migration of the young to distant places for jobs whilst the elderly remain behind alone in their homes. The number of crimes relating to them has increased. People lead a selfish life and there is a thorough disregard for older people. As a result of liberalization, privatization and globalization the consumption pattern is changing. Joint family is disintegrating, so there is no support for the grandparents. Loneliness sand frustration are making matters worse. The daughter-in-law has become too busy to provide adequate help to the elderly. As age increases, old people's status declines, their independence is curtailed and they lose self-esteem. Love, care and attention are what the elderly need most. Mohanan (2003) found that all elderly people in her study experienced one or other form of abuse in the family as well as in the old-age home. The most common forms of abuse experienced by them were neglect, lack of attention and denial of freedom, food and health care. Gender and employment status of the aged have a relationship with abuse. Elderly males and pensioners are least subject to abuse. Greethanand (2004) revealed that for the 60–65-year-old age group abused inside the house, the majority of incidents occurred during the day. She found several reasons for abuse such as financial difficulties, low health status, decrease in family support, failing moral values, urbanization, modernization and heavy traffic. Geethanjali (2004) explained how widows resident in old-age homes become very lonely and lose their economic support. The most important reason for staying in an old-age home is that there is no-one to take care of them in their own homes. The different types of abuse which they experienced in the old-age homes were verbal abuse, disrespect, displacement and no care during sickness, and no visits by their children. Saritha (2004) in her study found the most common form of violence against elderly people was committed by the care-givers within or outside the family. The plight of tribal widows is more pitiable because after the death of the spouse they depend on children and as such lose their status in the family. The elderly tribal widows face different types of abuse. The abuse is mainly from members of their own family, whereas they get more support from the members of their community. Soneja (2000), using the focus group discussion technique, found respondents aged 58–60 from higher and lower strata of society suffering abuse reported emotional problems, lack of emotional support, neglect, loss of dignity, maltreatment and disrespect by family members. Bambawale (2004) identified several abuses such as verbal abuse, emotional abuse, psychological abuse, exploitation, harassment, ill-treatment, deprivation and neglect. The majority of respondents reported that they didn't get respect from their family members and society. Some 25% of respondents mentioned that family members did not devote adequate time for them; 36.6% had tension related to the generation gap; 25% were stressed by their children's misdeeds. Kumar (2000), in a study on people aged over 60, found that 35% suffered from physical and psychological distress, Prakash (1999) found loneliness (21.05%) and neglect (27.3%) in a study in Udaipur.

Balambal (2004) has observed that abuse, neglect and violence could be seen in all the three strata of society, varying in each one. He found that elder abuse takes place within the four walls of the family and not much about it leaked out due to certain obvious reasons. There is not much physical abuse in the upper strata. But among the lower class, most elderly people are abused and neglected and face physical violence, too. The nature of abuse differs with different socio-economic cultural dimensions. As we go down the social ladder, the abuses are bold, open, ruthless and of the nature of personal intimidation. As we go higher up in the social ladder, more subtle abuses are evident. There exist young and old categories in society, with youngsters enjoying all the privileges of life while at the same time there is social denial for the old. Economic and social cultural aspects of society combine to push the elderly to the margins, and here lies the origin of abuse. Chakravorty (2004) in her Kolkata study on elder abuse found that poor elderly people are more vulnerable to physical abuse and abandonment, whereas psychological abuse, financial abuse and deprivation of love and care are experienced by elderly people belonging to the middle and higher classes. Pandey (2004) studied neglect and violation of elderly people in India by using primary and secondary methods. A large number of elderly people live with their offspring or close relatives. The abuse in most cases and in most of the incidents is habitual, chromic, verbal aggression and neglect. Physical violence is infrequent although not care, especially among the female elderly. Men are abused physically more than women when they live with their children.

Desertion is another form of abuse, as found in a study by Reddy (2004). It was found that family members had deserted 16% of elderly females, and desertion was more than three times higher among widows. As the elderly become old and disabled, the family members abuse or pay scant attention. Giri (2004) in a study on abuse, violence and neglect against elderly males revealed that a few of them lead solitary and destitute lives. They spend their remaining time in miserable conditions. An ICMR study in Tamil Nadu in 2002 revealed that the attitudes of family members towards older people varied: neutral (10%); rejected unwanted, just tolerated (38%); loved, respected but having no control over the family. Old people are sent to old-age homes even when families could take care of them, found Chandana (2004). Sometimes, children divide parents amongst themselves for providing care during old age, against the latter's will. Jamuna (1988) has found physical dependency, care-givers' stress and load, personality factor of care-givers, and inadequate support in family networks are major potential sources of elder abuse. Abuse amongst widows has been studied by many scholars, for example Jamuna (1988), Reddy (1992), Ramamurthi (1989). '[W]omanhood, widowhood and old age' are found to be risk factors for elder abuse. Abuse of the elderly has been studied amongst persons living with various diseases. Alzheimer's Disease sufferers are more vulnerable to mistreatment or abuse, says Malik (2000). Many entered old-age homes out of desperation at their treatment within the family. Most of the elderly people living in old-age homes suffer from one or other form of abuse: be it neglect, lack of attention, denial of freedom, food, health-care (Mohanan 2003). Bambawale in Pune found 40% of respondents felt totally neglected and the elderly women interpreted this as abuse. Neglect was felt due to non-inclusion of elderly people in important family matters and also less inter-personal communication among family members and the elderly.

In brief, from the review of literature it becomes evident that the abuse may vary from mild to severe; and it may be specific to culture and class. Abuse may be direct and indirect. Indirect abuse generally hurts elderly people emotionally and psychologically; and its occurrence appears as a common phenomenon. Direct abuse may occur rarely, depending on varied situations of family life. Judicial abuse occurs when elderly people's legal cases are dragged on for long periods, to the extent that the old person dies before justice is delivered. 'Abuse' looks to be a generic term. It refers to behaviours towards elderly people which: deprive them of essential basic necessities; force them against their genuine desires and needs; assault them; ignore them; neglect and marginalize them; and expect them to do what they can't do (Khan 2004).

Elderly people, with or without a spouse, living alone are a sociological reality of modern life. The number of such people will multiply enormously. The great challenges for them include:

- How to provide security?
- How to safeguard them from robbers, murderers and thieves, or even from some family or family-related persons who harm them in a disguised form?
- How can health care be provided when they are suffering with diseases and multiple disabilities?
- How to find people to look after them when family members are not available?

Ignoring these issues on the part of family, society, and country and leaving elderly people to live the life of highest insecurity and disease in home life are most severe form of abuse. But who is really abuser is a main issue (Khan 2004). All these give an impression that if the oldest old or very old are deprived of basic general care, health care, specialized health care, basic social security, food clothing, place of living and not provided services according to their body functioning, and if they are left uncared, unattended for their basic need by the care provider and if they are contrary condemned, criticized, rebuked, nagged, neglected, hurt abuse (Khan 2004). Depriving them of due rights, not creating and providing legal support to those suffering various atrocities committed at family and community level speaks of a social form of abuse which carries generic and specific meaning as well. Studies conducted by Khan (2006) and Khan and Smita (2006) have brought out different forms of elder abuse across different sections of society. In the lower class, the prominent forms of abuse are economic and physical, whereas in the upper class/strata of society the prominent abuse is emotional and social. The medical care abuse is almost the same in upper and middle classes. The medical care abuse in the lower class is not intentional. Through this study, authors have developed a scale for measuring the degree of abuse with a high degree of reliability and construct validity.

In the Indian context, most researchers in the area of ageing have tried to explore physical condition, social, emotional, economic and health conditions, health-seeking behaviours, inter-generation gap, attitudes towards the elderly etc. Research studies provide a very clear picture that the life of elderly people is full of problems of different kinds. And these problems are not individualistic, but rather deeply rooted into social systems, and transitions changing family structure, migration and mind-set. Over the years, elderly people in India have grown up without old age planning in the strong expectation that family would be the foundation of their old-age care. There is an absence of research into the dynamics of inter-personal relations among family members. When we try to understand elder abuse as a concept and as a social phenomenon, the situation is dice. Medical care, health care, health well-being, quality of inter-personal interaction, and elderly people's expectations of family members are absolutely crucial for understanding elder abuse and developing measures. Measuring elder abuse needs conceptual clarity and consensus which at the moment is missing.

Theorizing Elderly Abuse

Changing Sense of Worth Regarding the Elderly

An impersonal approach to life, unlike that in the past, has been accentuated by contemporary social, cultural, political and economic realities along with globalization. The dichotomy between individuality and collectivity is becoming more apparent. It appears that society as a whole, particularly in the urban areas, is becoming self-centric. The concern for others, as a glorious feature of Indian society, is decaying. This new development in society poses enormous danger at all levels. First, it affects the family as a potential unit of society; elderly persons suffer. India is a developing country, struggling with population problems, and with scarce resources which are insufficient to match the basic needs of a large section living below the poverty line. It overlooks the significance of "care for others" as a core value; most trivial conditions in the society are inevitable. Quite a lot of elderly persons are gradually being deprived from the nucleus of family care. There is a continuous breakdown of the family system, structurally and functionally. Such transitions are occurring under the influence of a growing culture of consumerism. The transition is not congenial to the care of elderly persons, among whom more vulnerability will emerge in the future. Tim Dyson, Professor of Population Studies at the London School of Economics has said that demographic expansion will pose many serious challenges.

Ancient literature in India is replete with reverent references to the elderly. Long life was cherished, old age was viewed with deference and the elderly used to play an important role as advisors and counsellors; and the family and community looked after them regardless of their productive capacity. In the twenty-first century, however, all the positive values attached to the elderly people and their care are bound to change since the social, cultural and family value systems in which elderly care values are embedded are in the process of fast erosion. Depleting emotional bonds in the new interaction pattern within the family, abysmal provision of health-care facility for the elderly, economic insecurity, growing scarcity of care-givers—all these provide clues that modern Indian society is heading towards vulnerability as far as the care of elderly people is concerned.

Growing Scarcity of Care-Givers

The scenario of care-givers is changing due to the rapid migration of young people, who are leaving the majority of older persons at the mercy of neighbours and in a culture of money order economy (MOE). Changing roles and expectations of women, their concepts of privacy, career, ambition and employment outside home imply considerably reduced time for care-giving. Millions of older persons are left in the rural areas without essential care (Khan 2001).

Imbalance Between Rural and Urban Areas

The majority of the population is shifting from rural to urban areas. The percentage of population in rural areas has declined from 82.75% in 1950 and is expected to come closer to 46.40% in 2025. Over the same period, the urban population has increased from 17.25% in 1950 to a projected 53.6% in 2025. Will the urban population maintain essential health facilities to keep its balance? Or will it tilt towards a catastrophic situation? Other features of this transition are that the country will have more literate elders than illiterate ones in years to come, and more females than men. The demand profile of needs will be different and more difficult than what we find today.

Scarcity of Space and Needs

Those who migrated to the larger cities years ago and settled down have now entered old age. The limited space in the house and increased number of family members over the decades have given rise to a new need pattern (NNP) amongst elderly people. The need could be broadly classified into six categories: nutritional; psychological; social needs; needs of special groups; economic needs; health; safety and security; and spiritual. All these areas represent a broader category, under which headings there are many further needs to be taken up for the care of elderly people. The perception of the elderly regarding their contribution to society's living arrangement, family life, programmes, and expectations from the family, society and government and the life preparatory measures are also vital reflection of needs.

Other needs are: protection against outliving one's savings; protection against being financially devastated by medical and other emergencies; protection against inflation; development and maintenance of a social support network; provision of opportunities for increased productive activity in a variety of ways; provision of opportunities for social and economic interaction in community control over some indoor and outdoor activities; and protection against illness and being physically assaulted. These are likely to be major factors in determining the quality of elderly care. The need profile of elderly people would also include addressing insecure feelings, counselling, consultation with a specialist, care by an attendant, physical aids, recreational needs, participation in cultural events, interaction inside and outside home, participation in religions activities outside home, desire to make a contribution to society, need to get respect, recognition, social support, and awareness about saving schemes and nutritional need (Khan 2001). Healthy ageing will depend a lot on meeting these needs.

Denial of the Naturalness of Ageing

Age-specific behavioural dispositions are primarily natural and universal. But the perception of such dispositions is culturally and socially learnt. Individuals' dispositions are always judged by society, and the individual learns, relearns and unlearns within the frame of society's normative behaviours. In the process of learning and unlearning, individuals generally extinguish those behaviours that do not serve their motive for gain and reinforce those that serve them. The behavioural disposition of children, youths and old people in society is determined by the gain and loss continuum. Why do societies deny the vulnerability of natural ageing and regard it as burden, unlike the vulnerability of the child? Why is the naturalness of ageing not construed in its real perspective? Why is the truth of ageing denied for elderly people? Why can the elderly themselves not take up ageing as a natural part of life, and nurture the art of living in proximity to naturalness? Why cannot society engineer and design systems that facilitate the process of healthy ageing? Why cannot economist calculate the economic requirement of vulnerable sections of society in order to address the issue of elderly care by family, community and government?

Although there is not authentic data revealing changes in the dynamics of family interaction, their attitudes and behaviour towards each other, nevertheless, scanning national newspapers, magazines and based on impressions derived from discussion with people, it appears that the crisis of old age is not only a demographic crisis but a value crisis, also. Inter-generational studies reflect on this issue. The influence of market forces and consumerism encroach upon society and gradually affect people regardless of their social, economic and cultural belongingness (Khan 2004). Elderly people across all sections of society are facing problems of disrespect, indifference and other emotional problems, though the nature of their problems may not be the same.

Transition in the Family

The joint family as a traditional care-taker of older people in India is a dying institution. The argument and advocacy to design quality care of elderly on the shoulders of the joint family is as hopeless as working against the tide. The nuclear family is the reality of the twenty-first century, and this too will further break into the 'micro nuclear family' in years to come as husband and wife reside separately due to employment opportunities. In fact, the micro family is already with us (Khan 2004). The pace and direction of developmental forces under this model, which primarily saturates the satisfaction of material needs, cannot be altered. So other basic needs—such as social, emotional, loving and caring ones—are gradually tilting towards a deprived condition for elderly care. The major changes in the future will be scarcity of family care (SCOFCA) for elderly people because of fast-growing globalization and industrialization, and it will accelerate the migration of earning members resulting in a peculiar situation whereby elderly persons are left alone. Scarcity in the migrated-from place, both physically and psychologically, will be a common phenomenon. The spell of ongoing transition in the institution of the family is being experienced in every section of society. The most visible impact of such inevitable transition due to globalization, industrialization and modernization has come in the life of older persons. The problem of isolation, alienation and loneliness leading to depression amongst these people is becoming more perceptible. Different forms of elder abuse are surfacing nowadays. Old-age insecurity is a new dimension for law enforcement agencies.

The ingredients of the traditional family system are changing fast, and perhaps the process cannot be reversed. So, nuclear and micro nuclear families (micro-family) are going to be the reality of modernization and globalization in the entire continent of Asia. In view of the rapid changes in the family, structurally and functionally, the need for institutional care is becoming more prominent in the big cities, particularly in cosmopolitan ones, which are relatively more urbanized and greatly affected by consumers culture and various market forces. The joint family as the best place of elderly cares looks like an outdated reality closely linked to the older values and morality of concern for the care and coordination of others. It is a major driving force of human behaviour. But such values are challenged under new emerging forces in society. Based on some inherent changes in the dynamics of relationships, the family could be classified into 'traditional family' and 'modern family'. These are somewhat like joint and nuclear family but not the same. Because psychological changes among the people towards the care of and concern for each other are occurring in both joint as well as nuclear families. Comparing the special features of these families can reveal the subtle differences across different dimensions; and these are very important for studying the basis of elder abuse.

Traditional Family In the traditional family, all family members don't assert themselves equally. Sometime one member is identified and recognized as the authority for taking decisions. As a head of the house, he/she commands authority in the family. Greater responsibility revolves around authority. The slightest favouritism shown to any member by the authority is likely to vitiate the emotional environment. Individual freedom and choice get least preference. Freedom for a free style of living is completely restricted. There is an absence of free-flowing emotions. Adjustment as a basic feature in the traditional family is most essential. Mutual care and concern are essential. Adherence to the family traditions is stronger, joint kitchen; joint celebration and joint action are highly desirable. An argument on every matter is not desired. These characteristics along with others (which are not listed here) are the main features of the traditional family.

Modern Family In the modern family, everybody asserts themselves equally and justifies individual standing on several family issues. No-one is recognized as the authority. Individual freedom and choice are key characteristics. There used to be a tendency of imposing authority on each other. It hampers the essence of adjustment. Highest priority is given to self-care, self-concern and an individualized style of living. Sharing with each other does not get due recognition. There is a complete flow of emotions, full argument and fewer of the consensus characteristics of the traditional family because of ongoing crises in the value system. Spontaneity in the expression of emotions without caring for the feelings of others is becoming a very common feature in society. It is quite possible that the members living in the joint family possess some important characteristics of the modern family and vice versa. But they may depend more on some features of the modern family. It appears that structural changes in the family per say are perhaps a secondary issue. The main issue is whether the nuclear family encourages some of the healthy features of the traditional family? Or does it completely embrace the value of the modern family. The care of older people may remain intact if some of the healthy characteristics of both survive as a value in the family. This issue requires future research to develop some suitable plan of intervention.

Productivity in Old Age

Perceptions of positivity and negativity towards old age are deeply embedded in productivity. In every society, individuals from childhood to old age are defined within a frame of productiveness of older persons in present as well as in future. For example, children's productivity is seen in career building to ensure productivity in future. The expectations and hopes for them are set out accordingly in the family. The moment hopes are lost and expectations are shattered, the definition about individuals also changes. And sometime the parents disown some grown-up children when their hopes and expectations are lost and vice versa. Quite often, family breakdown takes place due to such reasons and grown-up children (e.g. married or unmarried) opt for separate living. Sometimes, sons themselves disown their parents and demand partisan distribution of property as per social and legal practices prevailing in the society. The phenomenon of productivity, therefore, is a key to determining the status of the individuals, group and society. For example, a lot of negligence and discriminatory practices are followed in the case of girl child. This too appears to be closely associated with productivity and non-productivity. Discriminatory treatments are followed simply out of feeling of loss in productivity. The girls are generally perceived as a 'source of draining the resources' of the family. In other words, the family and society do not perceive the productivity of girls. Loss of expectations and productivity in society not only reinforces the psychology of indifference; it rather aggravates indifferent behaviours, and unwanted social evils such as family violence, gender bias, brutality towards the elderly, neglected care of older persons and female feticide etc.

Managing Loss of Productivity

Young old and old old classification fully justifies loss in productivity, which cuts the happiness of older persons. It needs serious discussion and we have to think how much the issue of loss can be successfully managed. Loss of productivity (LOPRO) in society cuts the happiness of older persons. Can elderly people's potential to remain productive up to the last breath of life be realized and recognized? The answer is yes, it is possible to mould the perception of society towards older persons. Society could be prepared to accept old age as a distinctive by older people is as an indicator of productivity. The characteristics of old age need to be positively reciprocated in the society. It would require multipronged strategies to the build futuristic policy and programmed.

Taking the above challenges into account, it looks as if the elderly themselves have to fight their own battle and win the war of ageing by establishing a new model of healthy ageing. They have to liberate themselves from the concept of negativity and non-productivity, and take up a proactive role with a new mind-set. They need to realize that they are capable of managing themselves and their surroundings provided they develop the capacity of knowing themselves and accepting their hidden potentiality. The issues of how 'they are old' or how 'they are made an old person' need to be carefully examined in the context of the global reality of older people (Khan 2011).

In every society, individuals over three main stages of life (i.e. child, youth and old person) are driven socially and culturally to define themselves distinctly and differently. As a child, the individual is expected to bubble with positivity and hope for a brighter future. As a youth, he or she is expected to be productive and earn resources. However, when that same individual becomes old, he/she is perceived as a symbol of negativity. This denotes a complexity of relationships and reciprocity amongst family members which is mainly determined by societal expectations. It also speaks to selfishness as an inherent characteristic of the society which values positively those whom it perceives as productive either in the present or in the future, and views negatively those whom it perceives as non-productive. Due to this. the image of old age in society is portrayed as a burden, helpless, dependent, vulnerable powerless, passive, pessimistic and depressive etc., suffering with a host of emotional conflicts, physical and psychological problems. Negativity associated with old age abuse as such is a result of social definition; that is, the way society images elderly people. As Julia Tavarez de Alvarez (1996) said, 'People in youth oriented societies perceive aging as a mitigated process of decline. People think that old age is necessarily a story of illness, poverty, desolation, and depression. Such attitudes lead people to accept as inevitable circumscribed roles in later life'. The crux of healthy ageing lies more in the transformation of images of old age in the light of longevity and increasing involvement of young old people into productivity.

References

Aber, J.L., and E. Zigler. 1981. Developmental consideration in definition of child maltreatment. *New Direction of Child Development* 11: 1–29.

Baker, A.A. 1975. Granny-battering. *Modern Geriatrics* 5: 20–24.

Balambal, V. 2004. *Study on abuse, neglect and violence factors in the threes strata of society.* Presented in national seminar on elderly abuse, Thiruvananthapuram.

Bambawale, U. 2004. *Abuse of the elderly: A myth or reality.* Handbook of Gerontology. Serials Publications. New Delhi, 243–267.

Burston, G.P. 1975. Granny Battery. *British Medical Journal* 3: 5–92.

Chakravarty, I. 2004. *Tackling elder abuse: Role of a voluntary organization in the city of Kolkata.* Presented in national seminar on elder abuse, Thiruvananthapuram.

Chandan, Jayesh. 2004. Old age: An age before death. www.oldage/buzzlr.com.

Geethanjali, R. 2004. Department of women's studies, Sri Padmavati Mahila Viswavidyalayam. Tirupati—Violence against elderly—A study on the female inmates of the old age homes. Presented in NS on elderly abuse, Thiruvananthapuram.

Giri, C. 2004. Violence against elderly male: A case study of Margherita, presented in Ns on elderly abuse, Thiruvananthapuram.

Gokhale, S.C. 2006. Roots of elder abuse part 1. Pune: ILC, April 20–23.

Greethanand, C. 2004. A content analysis of neglect and violence against elderly. Paper presented in NS on elderly abuse.

Jamuna, D. 1988. Study on Adjustment among pre and postmenopausal rural women. *Journal of Psychological Researches* 31: 88–92.

Johnson, T. 1995. *Wider mistreatment: Deciding who is at risk.* Westport: Connecticut Greenwood Press.

Khan, A.M. 2001. *Health care of elderly: Approach paper presented in consultative workshop.* Delhi: NIHFW.

Khan, A.M. 2006. Case studies of elderly abuse: Roots and bases intervention. *Journal of Indian Academy of Geriatrics* 2: 21–27.

Khan, A.M., and Smitha. 2006. Physical abuse of elderly in Indian context. *Indian Journal of Gerontology* 20: 121–130.

Khan, A.M. 2004. Decay in family dynamics of interaction, relation and communication as determinants of growing vulnerability amongst elderly. *Indian Journal of Gerontology* 2: 173–175.

Khan, A.M. 2011. Exploring older persons perception about old age and different form of elder abuse. *Help Age India Research and Development Journal* 17: 20–30.

Malik, M. 2000. Dynamics of caregiving-transition during three years prospective study. *The Gerontologist* 40: 185–198.

Mohanan. 2003. Abuse of the elderly in family and home for the aged. Paper presented in national seminar on elderly abuse.

Nayar PKB. 2004. Older Women in India: The context, issues and concerns (ed). Ashok Vihar, New Delhi: The Women Press.
Pandey, U.S. 2004. Dept. of Sociology, SPK PG college, Firozabad, Agra. Study on Neglect & Violation of elderly persons presented in National seminar of elderly abuse, Thiruvananthapuram.
Prakash, R. 1999. Study of morbidity pattern among geriatric population in urban area of Udaipur, Rajasthan. *Indian Journal of Community Medicine* 29: 1–3.
Reddy, K.N. 1992. Twilight years of life: Behavioural interventions. Paper presented in national seminar on "Healthy ageing—A life span perspective" at Department of Psychology Sri Venkeshwara University Tirupati.
Reddy, P.J. 2004. Department. of population studies, SU Uni., Tirupati. Study on neglect of rural elderly female in Chitoor, AP. Presented in NS on elderly abuse in Thiruvananthapuram, 2004.
Saritha, M. 2004. Elderly abuse—Study on elderly tribal widows. Paper presented in national seminar held Loyola College of Social Sciences, Thiruvanthrapuram.
Soneja, S. 2000. Elderly abuse in India—Study conducted for help age India & report submitted to WHO, (www.usd.edu/eabuse/archives).
Vadackumcherry, James. 2004. Nature, extent & intensity of violence on elderly-presented in National Seminar on elderly abuse, Thiruvananthapuram.
Wolf, R.S. 1996. Elder abuse mandatory reporting revisited. In *Violence, neglect, and elderly*, ed. L.E. Cebik, G.C. Graber, and F.H. Marsh, 155–170. Greenwich, CT: JAI Press.

Chapter 4
Family Violence: Reflections on Elder Abuse in India

Anupama Datta

Background

'Family violence' is an all-encompassing term that includes aggressive behaviour of members towards one another. The behaviour could be in the form of abuse, mistreatment, neglect suffered by the weaker, vulnerable and or dependent member; that is, sibling, spouse, child, parent. The victims are more often women, children and elderly. Child and wife abuse were the first to emerge in public discourse. The problem of elder abuse (which was initially called 'granny battering') was first talked about in a British journal in the 1970s. It is abuse perpetrated particularly by the care-giver/s with a risk or vulnerability model. So, it can happen in the home or in an institution. In the USA, this issue was viewed as a socio-political concern, and legislative action was taken. It is still a concern of the developed countries, though United Nations has recognized this as a growing menace and specified 'elimination of all forms of neglect, abuse and violence against older persons' as one of the objective of the Madrid International Plan of Action on Ageing (MIPAA) in 2002.

Family, as defined by sociologists, is the basic group where members are related to each other by birth or by marriage and mutuality is the guiding principle. This ideal family has undergone changes due to postponement of partnering and parenting; and an increasing trend, in many countries, of cohabiting unions which are often less stable than married ones. Increasing longevity and decreasing fertility have led to a movement from a 'bottom heavy' to a 'top heavy' family structure. The increase in life expectancy also leads to a remarkable increase in the years of shared lives across generations. This trend has significant implications for members of all generations,

The present article is based on the data collected by HelpAge India.

A. Datta (✉)
Policy Research and Development Department, HelpAge India, New Delhi, India
e-mail: anupama@helpageindia.org

Mala Kapur Shankardass and S. Irudaya Rajan (eds.), *Abuse and Neglect of the Elderly in India*, https://doi.org/10.1007/978-981-10-6116-5_4

who will share more years of life together than ever before in history. It will also affect the consensus on beliefs and attitudes, where the younger generations are more likely to subscribe to post-materialistic values like freedom of choice and self-fulfilment. This process of individualization is likely to result in less control of individual behaviour by family, community and religion. Religion, as we all know, has provided the moral values that cement family bonds and discourage anti-family behaviour. It is true that norms regarding filial obligations do not directly translate into care for the older members of the family, but they certainly provide the motivation or generate the intention to provide care.

Another related factor in this scenario of actual care provision is also conditioned by personal circumstances and the ‘opportunity structure’ including the children’s resources, whether they have competing obligations, and the extent to which there are alternative sources of help. Last but not least, the expectations and norms of the older parents themselves affect the inter-generational solidarity.

Elder abuse in the family, which is one of the most prevalent forms of violence against older persons, is a complex phenomenon which often has roots in multiple factors: family situations, care-giver issues, and cultural aspects. These factors may support, accept or deter violence against older persons. In traditional societies, for example, most of the older persons live in joint-families and, by common perception, still enjoy a high degree of integration in terms of the level of contact between members of different generations, the degree of positive feelings, the degree of consensus in beliefs and attitudes, the exchanged help, the norms of ‘family-ism’ and geographical proximity. However, three core aspects of solidarity (i.e. strength, nature and the direction of solidarity) may be under threat due to fast-paced changes in demography and ethos.

In the family, elder abuse may stem from: previous marital violence persisting into old age (spousal abuse); a continuation of violent relations between generations that existed in the family over many years; resistance/retaliation of abused when the older person is frail and dependent; stress at overcrowded living arrangements with little personal space; adult children being financially dependent on an ageing parent; economic hardships where an elder is considered an additional burden; an older person with physical and cognitive limitations leading to adjustment problems; care needs of the older member increasing beyond the coping ability of the family, to the extent that care-givers find themselves in an unfamiliar situation and unable to provide appropriate care.

Cultural factors associated with elder abuse include: the general devaluation of older people as they are regarded as ‘tired and retired’; lack of respect for age; considering home as a private domain and conversely keeping ‘family matters’ within the four walls; immigrants facing the challenges of unfamiliar language and system to report abuse; treatment of older women, particularly of widows who some countries may be branded as ‘witches’.

The characteristics of the abuser and abused are also important factors in violence against the elderly. The care-giver’s personality and stress could also result in mistreatment of the older person/s. Physical and or emotional problem of the care-giver; substance abuse; financial dependence of care-giver on older person; aggressive personality; low self-esteem of the care-giver; care-giver’s frustration at his/her unfulfilled needs; care-giver’s lack of support/resources.

The problem of elder abuse has become a worldwide phenomenon, but most of the research and discussion have been conducted in developed countries, and there is no universal terminology. There are currently three dominant theoretical frameworks: social gerontology; older person protection informed by geriatrics; intimate partner violence informed by the domestic violence movement. The differing understandings of these frameworks have resulted in differing research findings and prevention and intervention methods.

The most commonly accepted definition of family violence against elderly people is the one informed by social gerontology and sociology. The underlying theoretical frameworks include social exchange, ecological theory, care-giver's stress, and cycle of violence and abuser's impairment. The definition was proposed in the Toronto Declaration on Elder Mistreatment. It was adopted in the year 2002 at a meeting of the International Expert Group on Elder Abuse from the International Network for the Prevention of Elder Abuse (INPEA) and the World Health Organization (WHO) in Toronto.

Elder abuse is defined as '… a single or repeated act, or lack of appropriate action, occurring within any relationship where there is an expectation of trust which causes harm or distress to an older person'. It can be of various forms: physical, psychological, emotional, sexual, and financial or simply reflect intentional or unintentional neglect.

Elders in the family may face: physical infliction of pain or injury (common forms are beating, slapping, pushing, solitary confinement), threat, intimidation, coercion, fear of abusive behaviour or psychological abuse, verbal abuse, financial exploitation, fraud or coercion, undesirable sexual contact, neglect (intentional or unintentional denials of genuine care needs) and abandonment. The degree of severity may vary and have harmful consequences on the health and well-being of the older person.

It is difficult to estimate the prevalence of family violence against the older persons, mainly due to under or non-reporting by the victims and also differing understanding and definitions. According to various estimates, between 2 and 10% of older persons face abuse. In almost all cases, the perpetrator is a family member and in the majority of cases it is either an adult child or spouse.

Advanced age, gender, marital status, social isolation, deteriorating health and increasing dependence (physical/financial) on other family members are some of the risk factors. In other words, the oldest old, women particularly widows, lonely, poor, sick and dependent are more vulnerable to family violence. Depending on these risk factors the type and degree of violence may vary; for example, advanced age makes victims vulnerable to neglect and/or financial abuse; compared to men, older women are more vulnerable to physical and psychological abuse.

It is difficult to uncover this form of violence as there is worldwide reticence about it for various cultural and practical reasons: the older victims of family violence are dependent on the abusers, financially and/or emotionally; there may be cultural inhibitions in exposing the spouse or adult child's abusive behaviour; victims often lack viable alternatives forces; the victims continue to be in the relationship/house and desist from reporting. In many cases, victims start to blame themselves for such mistreatment.

Family violence against the older person exists and is probably on the increase due to changing demographic structure, longevity, structure and functions of the family, individualistic ethos, immigration and may related factors. Different theoretical frameworks have different understandings of the issue and also recommend different responses. However, the bottom-line is that something should be done to protect the vulnerable.

The second review and appraisal of the MIPAA in 2012 showed clearly that neglect, abuse and violence against older persons was acknowledged as a major policy issue in all regions, regardless of level of development. In the same year, the UN's General Assembly pronounced 15 June as World Elder Abuse Awareness Day, which has also become a main focus of the ongoing discussions of the General Assembly Open-Ended Working Group on Ageing.

For prevention, universal and/or selective prevention strategies are adopted. The former address the concerns of the general target group and the latter involve specific strategies for the various sub-groups that are at high risk. National and state governments, professional organizations, agencies and local voluntary groups have developed innovative prevention programmes. Services like health, social services, criminal justice and housing have been sensitized towards responding to the issues. Some programmes and initiatives may target older people as part of a larger campaign to prevent family violence, campaign against ageism, protect widows and single women, and promote community health. They may also mandate the training of workers in health centres to screen cases of abuse and promote the understanding and prevention of neglect and abuse of older persons. In some countries there is provision for mandatory reporting of such cases.

In the year 2006, INPEA designated 15 June as World Elder Abuse Awareness Day, and in 2010 this was ratified as a United Nations Day by the General Assembly. Governmental and non-governmental organizations (NGOs) around the world use this as an opportunity to educate and raise awareness about violence against older persons. In some countries, social service agencies register reported cases of elder abuse, conduct training, provide information and advice to community residents, and train professionals and amateurs to increase awareness and expertise.

Most of the developed countries of the world have laws against family violence and the concerns of older persons are subsumed under it. There are no specific provisions for the older persons in those laws, nor are there separate laws. In many developing countries of Asia and Africa there are no laws even for dealing with family violence. In some African countries, archaic laws criminalize witchcraft and generally older women are accused of it.

If family violence against older persons is assumed to be a significant social problem in future, the first step is to develop a sound understanding of its nature, causes and prevalence. It is important to conduct academic research to unravel the complex issue in its social and cultural context. It is also important to look at how it affects the most disadvantaged and marginalized older persons such as the disabled, widows, single women, rural elderly people, oldest-old persons etc. It is also important to understand the issue in less developed and developing countries so that appropriate responses can be developed.

Situation in India

The traditional norms and values of Indian society laid stress on showing respect and caring for the elderly. Consequently, elderly members of the family have normally been taken care of in the family itself. The family, commonly the joint-family type and related social networks have traditionally fulfilled the needs of the elderly in respect to social, psychological and economic security. In turn the elderly have known to contribute by dispensing their acquired wisdom, distributing their wealth, and keeping the members of the family in union. In other words, this relationship has been one of symbiosis and reciprocity.

Such a system of mutual support, however, is facing hardships in a modern, industrial society. The elderly have to depend more on their own resources than on group resources to fulfil their needs. Gore (1993) opined that in developed countries, population ageing resulted in a substantial shift in emphasis between social programmes, causing a significant change in the share of social programmes going to older age groups. But, in developing countries, these transfers take place informally and are accompanied by high social and psychological costs by way of intra-familial misunderstandings and strife. Assessing the impact of forthcoming social change on the status of elderly people, D'Souza and Fernandes (1982) apprehends that transition from agrarian to urbanized and industrialized society, from joint family type to unitary family type, and increasing emphasis on the individual as a unit would certainly deteriorate the position of elderly people. It is on account of these changes that it can no longer be assumed that our elderly live comfortably at home receiving care from family members willing to spare time and perform services for them. Given this trend, the elderly face a number of problems to adjust to in varying degrees. These problems range from absence of ensured and sufficient income to support themselves and their dependents, to ill-health, absence of social security, loss of social role and recognition, to the non-availability of opportunities for creative use of free time.

In this context of changing demographic scene and social values, the situation of the elderly in Indian society is becoming more critical, and of late abuse and neglect of the elderly in the family context is emerging as a significant problem. In a recent community study in a rural setting, an overall abuse rate of 48.2% was reported with physical abuse at 9% during the preceding year (Devi Prasad et al. 2009).

The research on ageing in India up to now has primarily been focused on socio-economic and demographic profiles, inter-personal relationships, problems related to retirement etc. especially of the urban elderly. However, certain issues related to the problems of elderly people have received insufficient attention from scholars, and one such area is elder abuse.

Elder Abuse in India

According to Devi Prasad, elder abuse warrants attention for three reasons: (a) seniors constitute a vulnerable group; (b) elder abuse, though not widely acknowledged, exists; (c) the maltreatment of the elderly significantly undermines their quality of life. Though there has been little work conducted in quantification of the extent of elder abuse, studies have sought to establish a ratio; for example, Rao (1995) looked at three villages in Andhra Pradesh and established that 40 out of 1000 elderly people had faced physical violence.

A review of the few studies focused on elder abuse indicates that the most likely victim of this is a female of very advanced age, role-less, functionally impaired, lonely and living at home with someone, primarily their adult child, spouse or other relatives. Studies in India have indicated that more women than men complain of maltreatment in terms of both physical and verbal abuse. The prevalent patterns of elder abuse mainly include: psychological abuse in terms of verbal assaults, threats and fear of isolation; physical violence; and financial exploitation. The health profile of the elderly victims indicates that a person suffering from physical or mental impairment and dependent on the care-takers for most of his or her daily needs is more likely to be the victim of elder abuse.

Older individuals suffering from depression, poor health or physical impairments were more at risk of being abused than those of similar age and normal health status. This indicates that a dependent elderly person with physical or mental impairments may be perceived as a burden by the care-givers, and the resulting stressful situation may more likely lead to the abuse and neglect of the elderly person. Though a large section of victims of elder abuse are less educated and have no income of their own, old people with high educational backgrounds and sufficient income are also found to be subject to abuse.

Coming to the profile of the abuser, it was reported that the son and daughter-in-law together and the spouse were mentioned by the elderly respondents as the most likely abusers (Siva Raju 2011). Besides the dependent position of the older person as a risk factor, other factors such as perceived powerlessness, social isolation, drug or alcohol addiction and anti-social behaviour of the abusers were also related to elder abuse. Certain major and frequently cited explanations about elder abuse are: cycle of abuse or inter-generational transmission of violence, dependence because of impairments, intra-individual dynamics, stress, negative attitudes towards the elderly and social-isolation. It is likely that abuse may occur through the inter-play of several of these factors. However, as mentioned earlier, studies conducted on abuse of the elderly have not been as extensive as desired. As Devi Prasad (2000) reminds us, 'there is little information either by way of empirical data or official statistics on the phenomenon of elder abuse in the Indian context. Even speculation about the nature of the problem has been difficult because of invisibility and sensitiveness of the problem'. Furthermore, the studies on abuse of the elderly conducted so far are related more to the Western world.

The present study is an effort to explore a contentious and inadequately studied issue in India, seeking to contribute to lessening the void in research that currently exists in the area of elder abuse. The study is expected to facilitate in evolving a suitable methodology to study the extent of abuse and forms of abuse faced by the urban elderly.

Though other problems faced by the elderly are also of critical importance, the incidence of elder abuse is highly associated with modernization, industrialization and other such processes of social change. With phenomena such as urbanization increasing rapidly, and given the trends in the population structure, it can be expected that the problems of the elderly and especially abuse of the elderly will rise. A Mumbai-based NGO received 200 calls about elder abuse in 2 weeks after it made a newspaper appeal for information on this issue (Veedon 2001). Most complaints related to problems with people's residence and financial vulnerability. Seniors living in slums reported instances of physical, emotional and economic abuse (Vaswani 2001). An elders' helpline in Bengaluru routinely receives calls from older people about harassment they face from their tenants, from noisy neighbours, and from cruel children (http://www.bcp.gov.in/Elders_Helpline.aspx). The Nightingale Trust reported having received about 700 calls in just eight months about harassment and abuse from older persons in the city of Bangalore (http://timesofindia.indiatimes.com/life-style/relationships/man-woman/Old-abused-and-harassed-the-story-of-Bangalores-elderly/articleshow/25261448.cms).

The incidence of elder abuse is empirically associated with modernization, industrialization and other such processes of social change. With these phenomena increasing, it is safe to expect that the problems of elder abuse will increase. A look at the number of calls received at HelpAge India's helplines across the country in the year 2012–2013 gives some weight to this apprehension (Table 4.1).

Studies Conducted by HelpAge India

In order to explore the reality of elder abuse in our country, HelpAge India has been conducting surveys on the subject from time to time. Different aspects and different cities were the focus of these studies to understand the reality. The current article is based on the results of the study conducted in the years 2013 and 2014. It was designed to explore the perception, experience, reporting and redress of cases of elder abuse in some selected Tier-I and Tier-II cities. In the year 2013, the selection of the 24 sample cities was based on the operations of HelpAge India's helplines. It included five metro cities and 19 Tier II cities. The list is given in Table 4.2.

Two-stage cluster sampling technique was used to select the sample for the research study. A commonly used two-stage cluster sampling scheme (the '30 × 7' sample developed by the World Health Organization) was used as reference. Instead, 20 × 40 formula was used; that is, 24 cities where HelpAge India runs

Table 4.1 City wise break up of calls received by HelpAge India's helpline in 2012–13

Sl. No.	Location	Total
1	Ahmedabad	975
2	Bangalore	1391
3	Bhopal	1071
4	Bhubaneswar	1322
5	Chandigarh	511
6	Chennai	2155
7	Cuddalore	3804
8	Dehradun	924
9	Delhi	1063
10	Eluru	2152
11	Goa	742
12	Guwahati	913
13	Hyderabad	2568
14	Jaipur	1033
15	Jammu	815
16	Kochi	517
17	Kolkata	2841
18	Lucknow	880
19	Mumbai	2269
20	Patna	1430
21	Pondicherry	2543
22	Shimla	321
23	Vijayawada	2628
	Total	34,868

elders' helplines were selected. A list of localities in each city was selected as the sampling frame. For each SEC, 10 localities were selected by systematic sampling with a random start. A sampling interval was obtained by dividing the total number of localities in a SEC by 10. The first locality was selected using a random number and then the second one after adding the interval and so on. In each locality, seven interviews—four men and three women—were carried out. The four men were selected following the right-hand rule of field movement from a landmark/starting point (e.g. school, temple, or some shop or any structure that was not a household) and three women were selected using the left-hand rule. In case of more than one eligible elderly person being in a household, the choice was made on the basis of availability and the youngest elderly person above 60 years.

In the year 2014, the study included male and female elderly people aged 60 or above, from all 12 cities across eight states. A sample of 100 elders per city was taken, with equal representation of males and females. Ten localities were selected by systematic sampling with a random start. The first locality was selected using a random number and then the second one after adding the interval and so on. In each locality, 10 interviews (five men and five women) were carried out. The five men

Table 4.2 List of cities included in HelpAge India's study on elder abuse in 2013

State	Selected metro city
Andhra Pradesh	Hyderabad
Delhi	Delhi
Maharashtra	Mumbai
Tamil Nadu	Chennai
West Bengal	Kolkata
State	Selected Tier-II city
Andhra Pradesh	Vizag
Assam	Sibsagar
Chhattisgarh	Raipur
Gujarat	Surat
Haryana	Faridabad
Himachal Pradesh	Solan
Jammu & Kashmir	Srinagar
Jharkhand	Jamshedpur
Karnataka	Mangalore
Kerala	Thiruvananthapurum
Maharashtra	Nagpur
Odisha	Cuttack
Punjab	Amritsar
Rajasthan	Bikaner
Tamil Nadu	Madurai
Tripura	Agartala
Uttar Pradesh	Kanpur
Uttarakhand	Haridwar
West Bengal	Durgapur

were selected following the right-hand rule of field movement (see above) and five women were selected using the left-hand rule. In the case of more than one eligible elderly person being in a household, the choice was made on the basis of availability and the youngest elderly person above 60 years. The states and cities are listed in Table 4.3.

In both years, an interview schedule with closed-ended questions was designed to collect data from the older persons to explore the reality of the following aspects of elder abuse:

- perception of what constitutes elder abuse
- perception of prevalence and most prevalent type of abuse
- personal experience of abuse
- older persons' reporting behaviour
- suggestions for effective measures to deal with elder abuse
- information about currently available intervention mechanisms.

Table 4.3 List of cities included in HelpAge India's study on elder abuse in 2014

Sl. No.	State	City
1	Karnataka	Bengaluru
2	Karnataka	Mangalore
3	Tamil Nadu	Chennai
4	Tamil Nadu	Madurai
5	Delhi	Delhi
6	Assam	Guwahati
7	Andhra Pradesh	Hyderabad
8	Andhra Pradesh	Vishakhapatnam
9	Uttar Pradesh	Kanpur
10	West Bengal	Kolkata
11	Maharashtra	Mumbai
12	Maharashtra	Nagpur

Socio-economic Profile of the Respondents

In the year 2013, the majority of the respondents (69%) were in the age group of young-old seniors (aged 61–69), 60% were currently married and 38% were widows/widowers. Some 83% lived with the family, 11% lived with the spouse only, and 6% lived alone. Some 73% owned the house in which they lived. Some 23% were retirees, 33% homemakers, 7% currently employed in private jobs, 10% labourers, 4% skilled labour, 4% small traders and 5% in business. Some 21% were illiterates, 53% had various levels of schooling from primary to senior secondary, 16% were graduates and 9% postgraduates.

In the year 2014, 79% of the respondents were in the age group of 60–69 years (i.e. young-old) and 21% in the age group of 70–79 years (i.e. old-old). The mean age of the elderly was 67 years for men and 66 years for women. Around 61% of the elderly were 'currently married' and living with their spouse, 28% were widows and 9% widowers. The percentage of currently married elders was higher among males (79%) than among females (43%). The percentage of widows was greater than that of widowers. About 26% of the elderly were illiterates, 5% graduates and 1% postgraduates. Around 26% had had up to primary education, 20% had middle education and 15% completed secondary education, while 11% completed higher secondary education. The literacy rate was higher among males. About 39% of the elderly were homemakers, 13% had retired from state or central-government service, 9% had retired from private companies and 13% were skilled/unskilled workers, while 3% were casual labourers. Around 62% of the elders reported that they owned their house, 31% reported that the house they resided in currently was rented, and 7% of the elderly reported that their current residence was owned by others. Some 77% of the elderly lived with their family compared to 14% who lived with a spouse only. Some 7% reported living alone.

Findings

Opinions of Elderly People on Elder Abuse

Disrespect, verbal abuse and neglect were identified as the major components of abuse as understood by elderly people in the context of elder abuse. Neglect as a sub-type of abuse was seen more among elderly people from Tier-II cities, while disrespect and verbal abuse were seen more in the Tier-I (Metro) cities.

Some 72% of seniors from Tier-I (Metro) cities and 71% from Tier-II cities opined that elder abuse was prevalent in our society. In Tier-I (Metro) cities, more males agreed with the statement regarding prevalence of elder abuse, while in Tier-II more females agreed. Out of those who agreed, 48% thought that it was moderately prevalent, while 31% thought that its prevalence was high and 21% thought it was low.

According to seniors, disrespect (76% in Tier-I [Metro] cities and 69% in Tier-II cities) was the most prevalent form of abuse followed by verbal abuse (61 and 72% respectively), and neglect (54 and 63% respectively). However in cities in Andhra Pradesh, Rajasthan, Maharashtra and Tamil Nadu a high percentage of seniors reported beating/slapping as a prevalent form of abuse.

In the year 2014 also, most seniors associated elder abuse with disrespect, verbal abuse and neglect. Verbal abuse ranked the highest, followed by neglect and disrespect respectively.

Some 83% of the elderly said that elder abuse was prevalent in society, only 17% of elderly felt that it does not exist. Out of those who opined that elder abuse was prevalent in society, 34% thought that the prevalence was high, while 56% thought that it was moderately prevalent and only 10% thought it had low prevalence. More elderly people in Tier-II cities compared to those in Tier-I cities thought that elder abuse was prevalent in society.

According to the elderly, verbal abuse (32%) was the most common form of abuse followed by disrespect (27%) and neglect (22%). In Kanpur, Mumbai, Visakhapatnam, Chennai and Madurai, higher percentages of seniors reported beating/slapping as a prevalent form of abuse. Economic exploitation was reported by many as a prevalent form of abuse in Mumbai, Visakhapatnam, Chennai and Bengaluru.

Personal Experience of Elder Abuse

Some 23% reported personal experience of elder abuse. Among the Tier-I cities Hyderabad and Kolkata reported rate of abuse as more than the national average at 37.5 and 28%. Delhi was closer to the national average with 20% followed by Mumbai (11%) and Chennai (10%). Among the Tier-II cities, Madurai recorded the highest incidence of elder abuse (63%) followed by Kanpur (60%). Amritsar reported the lowest experience of elder abuse, below 1%.

In Tier-I cities, 22% males compared to 29% females experienced abuse, while in Tier-II cities, 21.5% males compared to 25% females reported experiencing abuse. About 72% of those who experienced abuse belonged to the age group of 60–70 years while 19% belonged to the age group of 71–80 years.

In the year 2014, across the cities, 50% of the seniors admitted to personal experience of abuse. The above stated 50% is a quantum jump from the previous year's average of 23% admitting to personal experience of abuse. A higher percentage of females (53%) reported abuse, as against males (48%).

About 72% of those who experienced abuse were in the age group 60–69 years while 25% belonged to the 70–79 years group. Facing abuse was reported more by females (53%) than males (48%).

The actual experience of abuse was higher (50%) as opposed to the perception of 'High' prevalence of abuse at 34%. Interestingly, Kolkata and Bengaluru reported the highest 'Actual' 'Incidence' of elder abuse at 60 and 75% as against the 'Perceived' 'High' extent of elder abuse at only 38 and 9%. Further, Hyderabad rates at 9% for 'Perceived' 'High' extent of elder abuse as against 'Actual' Incidence of Abuse at 40%. Bengaluru had the highest percentage of elder abuse amongst Tier-I cities at 75%, Kolkata 60% and the least cases of abuse were reported from Delhi with just 22% of elderly facing abuse. In Tier-II cities, Nagpur had the highest prevalence of elder abuse (85%) followed by Mangalore (73%). Kanpur recorded the lowest at 13%.

Type of Abuse Experienced

On a national level, the three main types of abuse experienced were disrespect (79%), verbal abuse (76%) and neglect (69%). In Tier-I (Metro) cities, disrespect was the most common type of abuse faced by the elderly, followed by neglect (67%) and verbal abuse. In Tier-II cities, verbal abuse is the most common type of abuse (81%) followed by disrespect (75%) and neglect (66%). It is important to mention here that physical abuse like beating and slapping is also not so uncommon in the country. In Tier-II cities, Nagpur, Haridwar, Dugapur, Jamshedpur, Faridabad and Vizag reported fairly high incidence of beating/slapping. Comparatively speaking the incidence was lower in other cities like Raipur, Cuttak and Srinagar. Among Tier-I cities, Delhi reported fairly high incidence. Another point to note is that in most cities, this type of abuse was faced more by women than by men.

In the year 2014, as stated earlier, across the cities, verbal abuse (41%), disrespect (33%) and neglect (29%) were the major types of abuse faced by the elderly. Among the elderly who were abused, 34% from Kolkata faced disrespect, while in Guwahati and Visakhapatnam 38 and 49% respectively faced verbal abuse and 26% from Hyderabad faced neglect.

Frequency and Duration of Abuse

Among the elderly who faced abuse, 28% reported that they had done so for 3–5 years, while 26% reported the duration as 1–2 years. About 6% of the elderly had been facing abuse for more than 15 years. One-third of the elderly who faced abuse reported that they faced it almost daily (35%) and 23% faced it once a month. The frequency of abuse was generally more for older women than for men.

In Tier-I cities the highest average number of years of abuse (6.26) was reported from Mumbai, and the lowest (3.96) from Hyderabad. In all Tier-I cities except Mumbai, the average number of years of abuse for women were less than that of men. In Tier-II cities the average years of abusive experience of elderly people ranged from 1 year in Durgapur to 14 years in Bikaner. In eight cities in the Tier-II category, women suffered more than men in terms of duration of abuse. It is also important to mention here that in Amritsar and Durgapur, with low rates of experience of elder abuse, only women reported abuse to others.

In the year 2014, among the victims, 46% reported having faced abuse for 3–5 years, while 25% reported the duration as 1–2 years. About 4% of the elderly had been facing abuse for more than 15 years. The mean duration of abuse was 5.2 years. Compared to 2013, there was a sharp increase from 28 to 46% in the cases where duration of abuse had lasted 3–5 years. The highest number of cases of abuse for the longest duration interval of abuse (11–15 years) was reported from Hyderabad and Nagpur. Mumbai showed the highest incidence of 3–5 years.

Some 35% of the victims reported that they were abused at least ‘once a week’, 20% reported having faced abuse ‘once a month’ and 17% reportedly faced it ‘almost daily’. Seniors who faced abuse ‘almost daily’ in Tier-I cities were most numerous in Hyderabad (42%) and least numerous in Mumbai (26%). The number of those facing abuse ‘once a week’ was highest in Mumbai (47%) and lowest in Hyderabad (13%). The elders who faced abuse ‘almost daily’ in Tier-II cities were most numerous in Guwahati (71%) and those facing abuse ‘once a week’ were most numerous in Mangalore (99%).

Primary Perpetrator of Abuse

Nationally, the daughter-in-law was reported by 39% as the main perpetrator of abuse, very closely followed by the son (38%). Among the perpetrators of abuse in Tier-I (Metro) cities, 44% reported the daughter-in-law as the main perpetrator, while one-fifth reported the son (24%) and 23% reported relatives. In Tier-II cities, 33% reported the daughter-in-law as the main perpetrator, followed by the son (18%) and daughter (17%). It is also interesting to note that in Tier-I cities, in Chennai and Mumbai, fewer women victims blamed the son for abuse. There was not significant gender difference in those blaming the daughter-in-law except in

Delhi where fewer male victims blamed the daughter-in-law. Kolkata and Hyderabad did not show gender difference in reporting the abuser.

In Tier-II cities, the trend of fewer women blaming the son was visible in only four cities. There was no significant gender difference in nine cities. It is important to mention here that in Jamshedpur none of the abused blamed the son.

In the year 2014, seniors across cities were asked about the abusers within their family. The daughter-in-law (61%) and son (59%) emerged as the top perpetrators. This is a trend that is continuing from the previous years. Not surprisingly, 77% of those surveyed live with their families.

Amongst Tier-I cities, Mumbai reported the daughter-in-law as the main perpetrator of abuse (92%) while Hyderabad reported the daughter-in-law at the lowest level of 40%. The son as perpetrator was reported by 65% of abused elderly people in Bengaluru. and the lowest rates of 21% in Mumbai and 23% in Delhi. In Tier-II cities, the daughter-in-law was mostly reported as the perpetrator in Mangalore (97%) and least so in Vishakhapatnam (51%). The son was also mostly reported as a perpetrator in Mangalore (95%) and least so of in Kanpur (23%).

The son and daughter-in-law were reported as the main perpetrators of abuse in cases of elderly males and elderly females respectively. Across the 12 cities, 65% of elderly females and 57% of elderly males reported the daughter-in-law as the perpetrator while 60% of elderly males and 58% of elderly females reported their son as the perpetrator.

Reporting Behaviour

Among those who experienced abuse, 30% made an attempt to report it. Among those who faced abuse, 37% from Tier-I (Metro) cities made an attempt to report the matter, while 26% from the Tier-II cities did so. In Tier-I cities, only 10% of the victims from Kolkata reported abuse whereas 48% in Chennai did so. Hyderabad and Mumbai were lower than the national average and Delhi over the national average at 41%. In Tier-II cities Srinagar, Amritsar and Bikaner, none of the victims reported abuse and at the other end of the scale was Nagpur where 70% of the victims reporting abuse, followed closely by Durgapur (64%). In 10 cities including Nagpur and Durgapur, reporting of abuse was higher than the Tier-II city average.

In the year 2014, among those who experienced abuse, 41% did not report it to anyone. Reporting of abuse was highest in Mangalore (88%) and lowest in Kolkata (12%). Compared to the previous year, there was a substantial increase in the percentage of reported experiences of abuse from 23 to 50%, and a corresponding increase in the reporting pattern from 30 to 59%.

Conversely, the percentage of elderly 'not reporting' abuse decreased in all cities except in Nagpur, Kanpur and Delhi. In fact, in Nagpur there was a marked increase from 29 to 44%, whereas in Kanpur and Delhi there was a marginal increase.

However, Kolkata consistently recorded the highest percentage of non-reporting of abuse (90% in 2013 and 88% in 2014). The sharpest decline was recorded in Vishakhapatnam, from 73 to 19%.

Person/Agency Approached to Report Abuse

Nationally, 42% approached a relative, 30% approached other family members, while 28% approached a friend. Relatives have been approached by 37% of the elderly from Tier-I (Metro) cities and 50% from Tier-II cities to report the abuse they have faced. About 33% from Tier-I (Metro) cities and 27% from Tier-II cities reported having approached other family members, while 30% of the elderly from Tier-I (Metro) cities and 23% from Tier-II cities reported the matter to their friends.

It is interesting to note that in Tier-I cities, except for Chennai, besides the informal mechanisms, victims were using formal channel as well. Local police were approached by elderly women more than elderly men in Kolkata and Delhi. In Hyderabad and Kolkata, older women approached community leaders as well. In Tier-II cities, in Mangalore, 52% of the victims reported abuse and women doing so outnumbered men. But, they only used informal channels like family, friends and relatives. In the other cities in South India, both formal and informal agencies were used to report abuse, including NGOs, which were mentioned mostly by women. Reporting abuse in the Tier-II cities in North India was abysmally low and was mainly restricted to family, friends and relatives. It is important to mention here that in Solan only 2% reported abuse and all of these victims were women. In Faridabad, women reported only to informal agencies, whereas men reported to police as well. The only exception in the case of North Indian cities was Kanpur where 41% of victims reported it and women made attempts to report abuse to community leaders, NGOs and police. In cities in Western India, the rate of reporting was higher than the national average, and in Surat, as in Mangalore, women outnumbered men in reporting abuse. But, there was no significant gender difference in use of agencies for reporting abuse. In the cities in Eastern India, except in Raipur, the reporting was higher than the national average. But reporting was done mainly to family, friends and relatives. The only exception was Durgapur where more women than men were reporting abuse to the community leaders.

In the year 2014, 53% of the abused elderly people approached a relative, 42% approached their friends, while 35% approached either a family member or an extended family member to report the abuse. The highest reporting to a relative was seen in Kolkata and Guwahati. The highest percentages of reporting to other family members were in Kanpur and Nagpur. Approaching a family member to report abuse was reported the most in Kanpur (80%). About 13% from Delhi reported that they had approached an NGO to report abuse.

Reason for Approaching Any Person/Agency

The elderly people who faced abuse and approached a person or agency were asked about their reasons for doing so. 'Maintaining confidentiality of the family matter' emerged as the major reason (71%) followed by 'confidence in the ability of the person/agency to solve the problem' (62%). Half of the elderly people (51%) reported that they 'did not know any other way to deal with the problem' and hence they approached the agency/person. There were no significant city or gender differences.

Reason for Not Reporting Abuse Cases

The seniors who did not report about the abuse faced by them were asked about the reasons for the same. 'Maintaining confidentiality of the family matter' was the major reason behind not reporting abuse (31%) followed by 'fear of retaliation' (23%). One-fifth of the elderly people (21%) did not report it as they did not know how to deal with the problem. There were no significant city or gender differences.

Information About Currently Available Redress Mechanisms

There are few formal redress mechanisms available in our country to deal with elder abuse, and awareness is the first step to be able to use them effectively. The study found that Police Help Line (70%) was the most commonly known redress mechanism for elder abuse. Some 18% reported being aware of HelpAge India's Elder Helpline number, while 11% were aware of the Maintenance and Welfare of Parents and Senior Citizens Act, 2007. About 61% of seniors were aware of only one of the redress mechanism, while only 8% were aware of two redress mechanisms and 2% reported to know more than two redress mechanisms. Some 29% of the elderly were not aware of any redress mechanisms.

If we look at Tier-I cities we find that in Chennai and Hyderabad, more than 50% of the respondents were not aware of any formal reporting and redress mechanism. In Kolkata, 25% were in this category. In Delhi and Mumbai the proportion of such people was fairly low. It is interesting to note that in cities where the overall awareness was low (i.e. Hyderabad and Chennai), out of those who were aware of such formal mechanisms, a fairly high percentage (35 and 19% respectively) of older persons were aware of the Maintenance and Welfare of Parents and Senior Citizens' Act, 2007. The awareness about this Act was very low in other Tier-I cities. Comparatively fewer older women were aware of the Act than men in these cities.

In Tier-II cities, we find that in cities in South India, there was lack of awareness about any formal reporting and redress system in comparatively larger proportion of respondents, ranging from 20% in Thiruvananthapuram to 62% in Mangalore. The awareness about the Act ranged from a low of 8% in Mangalore to 23% in Madurai. In comparative terms, more men than women were aware of the Act except in Thiruvananthapuram. But women were equally or better informed about the police helplines and or other mechanisms.

In Tier-II cities in North India, we find that those in the category of older persons who lacked information about any formal reporting and redress system ranged from a low of 3% in Bikaner to a maximum of 46% in Faridabad. But it is interesting to note that awareness about the Act was abysmally low in these cities, older women faring worse than men on this count except in Bikaner, Haridwar and Solan. Police helplines were known to all and in many cities the gender difference was not significant. In Kanpur and Haridwar, comparatively more women were aware of the other formal mechanisms than men.

In Tier-II cities in Western India, there were more differences in trends than similarities. In Nagpur, 49% of the respondents were not aware of any formal reporting and redress mechanism whereas this was as low as 3% for Surat. In Nagpur, 12% knew about the Act compared to 18% in Surat. In Surat, comparatively more women than men were aware of the Act and vice versa in Nagpur. However, the police helpline was better known and gender difference was insignificant.

In Tier-II cities in Eastern India we find that the respondents who did not know about any formal reporting and redress system were 0% in Durgapur and Raipur. However, awareness about the Act was rather low in these cities; there was complete lack of awareness about the Act in Agartala and Raipur. The awareness about the Act was better in Cuttack, Jamshedpur and Sibsagar. Police helplines were better known mechanisms in almost all cities and the gender difference was not significant.

In the year 2014, a large majority of older persons (64%) were aware of the Police Helplines, whereas awareness about the other mechanisms was not as widespread. Only 14% were aware of the Maintenance Act and 9% about HelpAge India Elder Helplines. It is important to note that 18% were not aware of any mechanism.

Suggestions for Effective Measures to Deal with Elder Abuse

Sensitizing children and strengthening inter-generational bonding has been reported as the most effective way to deal with abuse by the elderly (58%). More than half (55%) of the elderly added that increasing economic independence of the abused and sensitizing young adults are effective ways to deal with the abuse.

In Tier-I cities, except for Mumbai, all cities showed preference for increasing economic independence of older persons to reduce elder abuse. The other top choices included developing legal and social systems to deal with abuse and sensitization of children and young adults. In Tier-II cities, in cities in South India, except for Thiruvananthapuram, the general preference was for increasing economic independence. Sensitizing children and young adults were the other common preferences. Besides these, only older persons in Thiruvananthapuram showed preference for developing effective legal reporting and redress mechanisms. In cities in North India, increasing economic independence was not a general preference in Solan and Kanpur, and showed the majority response to be in favour of sensitizing children and adults and developing effective legal response system. Older persons in Srinagar showed preference for developing effective social reporting and redress systems. In Bikaner, developing self-help groups for older persons was suggested as an effective way to deal with elder abuse. Sensitizing children and young adults was an almost universal choice. In the cities in Western India, the common preferences were developing an effective legal system to deal with elder abuse and increasing economic independence besides sensitization. In cities in Eastern India, developing an effective legal system was the preference of five out of six cities. Sensitizing children and young adults were the other almost universal choices in all the cities. It is interesting to note that in Jamshedpur, the overall preference was for developing self-help groups of older persons and an effective social reporting and redress mechanism to deal with the problem. Increasing economic independence was also articulated as one of the best ways to deal with elder abuse in four out of six cities in the region.

In the year 2014, there was not much difference in the opinion of the victims on this issue. The top-three-ranking effective measures identified in that year were similar to the ones identified in the previous year: 'increasing economic independence of the victim' and 'sensitizing the younger generation'. However, 'Developing an effective legal reporting and redress system' is one measure which does not appear in the national aggregate figure, but was pointed out as being an important step for effectively dealing with elder abuse by respondents in Bengaluru, Delhi, Mumbai, Hyderabad and Kolkata in Tier-I cities and by victims in Mangalore, Madurai and Kanpur in Tier-II cities.

Discussion

The study found that elder abuse was not unheard of and uncommon in Indian cities. The proportion of elderly respondents who had personal experience of abuse was fairly high. It is important to note that in 2013, out of nine cities that reported a higher rate of abuse than the national average, only two were Metro cities. Madurai and Kanpur, both Tier-II cities, topped the list with 63 and 60% respondents reporting personal experience of elder abuse. The trend continued in 2014 as well, when four out of six Tier-II cities reported more than the national average. It is also important to mention here that the cities that reported fewer than 5% cases of elder abuse were in Assam and Tripura. The experience of abuse was reported more by older women than men and this was true for both Tier-I and Tier-II cities.

Disrespect, verbal abuse and neglect were the most commonly reported forms of abuse; though physical abuse, like beating and or slapping, and economic abuse were also not so uncommon. Physical abuse was experienced more in Tier-II than in Tier-I cities in 2013 but was not reported by statistically significant numbers in the following year.

The intensity and duration of abuse reported by the victims also should be taken into account as many reported facing it every day; 26% of the victims reported facing it for 3–5 years, but in the year 2014 it increased to 46%. Here, also, the Tier-II cities fared worse than Tier-I cities. Gender differences were also reflected in this case. Older women suffered more than men.

The primary perpetrators of abuse were daughters-in-law and sons in Tier-I cities, but in Tier-II cities sons and daughters as well as daughters-in-law along with were to be perpetrators of abuse. Daughters and sons-in-law were also mentioned as abusers by the victims.

Elder abuse is not one of the world's most talked-about behaviours, and this is no less true in India where it is considered a private family matter. However, the study revealed that many of the victims reported it. It was reported more in Tier-I than Tier-II cities. In some Tier-II cities none of the victims reported abuse; though the reported rate of abuse was fairly high in these cities; whereas in some other cities' rates of abuse and reporting were both high.

Most of the victims of abuse chose to report the matter to family, friends and relatives, but it is interesting to note that formal channels of reporting and redress were also used by the elderly. In Tier-I cities, more elderly female victims in Delhi approached the local police than did elderly male victims. In Hyderabad, women also approached community leaders. In Kolkata, elderly female victims approached both local police and community leaders. In many Tier-II cities, senior women also approached NGOs, community leaders and the local police. In many Tier-II cities where the reporting of abuse was high, it was mainly seeking intervention of informal channels like family, friends and relatives. But, the significant point was that comparatively more women victims were reporting it than men.

Those elderly victims who did not report abuse were mainly influenced by a wish to protect family prestige and/or out of fear of retaliation. Many did not know how to deal with the challenge.

It is not surprising that the levels of awareness among the respondents about the formal mechanisms available to report and redress abuse were low even in Tier-I cities. Many older persons were not aware of any such mechanism. The awareness about the legislation that protects the older persons from abuse was abysmally low across the cities. Women in most cities lagged behind men in this awareness. However, in many cities women were equally aware about the police helpline and in some others more women were aware about the helplines and mechanisms developed by NGOs and community to deal with abuse.

The elderly in urban India thought that the best way to effectively reduce elder abuse was to sensitize children and young adults, and also to increase the economic independence of the older persons. However, in many cities their preference was for developing an effective legal system to deal with the menace.

The study busted many myths around elder abuse in the country and provided insights into many areas that need urgent attention. Elder abuse was not something that happens in big cities only—some of its worst manifestations were found in smaller cities. It affected both men and women, but women more than the men. However, women did not take it quietly in their homes and made efforts to report it informally and where possible to formal systems. Women were not shy about reporting it. Our system lacks in its ability to relay relevant information to the marginalized and vulnerable segments of society. Herculean efforts should be made to strengthen the information-dissemination system. Efforts should also be made to strengthen existing reporting and redress mechanisms and develop new ones so that these act as deterrents.

References

Bengaluru City Police. http://www.bcp.gov.in/Elders_Helpline.aspx. Accessed 8 Sept 2016.

Devi Prasad, B. 2000. Abuse and neglect: A review of research and programmes. In *Gerontological social work in India*, ed. M. Desai, and S. Raju. Delhi: BR Publishing Corporation.

Devi Prasad, B., Aruna Khasgiwala, and Thrity Vaswani. 2009. Families in difficult situations. *The Indian Journal of Social Work* 70 (2): 189–217

D'Souza, Alfred, and Walter Fernandes (eds.). 1982. *Aging in South Asia: Theoretical issues and policy implications*. New Delhi: Indian Social Institute.

Gore, M.S. 1993. The elderly in an ageing society, Contributed to Volume IV of Encyclopedia on Ageing, Japan.

HelpAge India. 2013. Elder abuse in India (2013): A HelpAge India report. New Delhi. Unpublished Report. https://www.helpageindia.org/images/pdf/ElderAbuseIndia13.pdf. Accessed 27 April 2017.

HelpAge India. 2014. Elder abuse in India (2014): A HelpAge India report. New Delhi. Unpublished Report. https://www.helpageindia.org/images/pdf/elderabuseindia14.pdf. Accessed 27 April 2017.

Nightingales Trust. http://timesofindia.indiatimes.com/life-style/relationships/man-woman/Old-abused-and-harassed-the-story-of-Bangalores-elderly/articleshow/25261448.cms. Accessed 8 Sept 2016.

Rao, K.V. 1995. Rural elderly in Andhra Pradesh: A study of their socio-demographic profile. *Unpublished Ph.D. Thesis*, Visakhapatnam, Andhra University.

Siva Raju, S. 2011. Studies on ageing in India: A review. BKPAI working paper No. 2, United Nations Population Fund (UNFPA), New Delhi, 12.

Vaswani, T.G. (2001). Family care of the elderly: Abuse, neglect and abandonment. *The Indian Journal of Social Work* 62 (3): 492–504.

Veedon, R. 2001. Elder abuse in the urban context. *Indian Journal of Social Work* 62: 480–491.

Chapter 5
Anatomy of Elder Abuse in the Indian Context

Renu A. Varughese and D. Jamuna

One of the remarkable features of the twenty-first century is the parabolic growth in the number of seniors in the total population. Other notable developments are the higher proportion of females, the bulge in the segment in the population aged over 80, and the increase in lifespan after the age of 60 almost five times higher than when India achieved independence. Demographic changes along with a series of socio-cultural and technological changes brought challenges not only for the growing elderly population but also to their families and society. Though population ageing is a topic of concern among gerontological scientists, it is a difficult proposition in the changing socio-cultural context.

Migration of the younger generation (care-providers) and dual-career families where women are transforming their roles as major economic contributors rather than being domestically oriented housewives are becoming major issues in elder care in the Indian context. In the absence of strategic measures to meet the challenges, the extended years of life will have no meaning for senior citizens. It is obvious to expect that well-defined and distinctive strategies for livelier longevity will strive to ensure respect, dignity and worth in old age without victimization, mistreatment and abuse at the familial and societal levels. The paradigm shift in the value system of the ancient agrarian society has resulted in disintegration, disorientation and collapse of the joint-family care and support systems (Jamuna 2004, 2006a, b). Consequently, seniors are forced to remain devoid of a support system at home or choose an option for alternative living arrangements. Yesteryear's practice

R.A. Varughese
Mercy College, New York, NY, USA
e-mail: rvarughese@mercy.edu

Present Address:
R.A. Varughese · D. Jamuna (✉)
Department of Psychology, Centre for Research on Ageing,
S. V. University, Tirupati 517502, India
e-mail: jxd172@gmail.com

Mala Kapur Shankardass and S. Irudaya Rajan (eds.), *Abuse and Neglect of the Elderly in India*, https://doi.org/10.1007/978-981-10-6116-5_5

of elder care as a dutiful task of a primary care-giver has been dwindling due to lack of time, commitment and dearth of resources for care giving. This has resulted in the emergence of adult day-care centres, retirement homes, and residential care facilities of all types in India. However, due to the prevailing stigma against institutionalization, and lack of acceptance for it in contemporary Indian society, families are forced to keep their older members at home in spite of the lack of time and resources for care giving (Jamuna 1995, 2006b). Increased demands of care giving and stress often push the care-giver to a state of physical and mental exhaustion that cannot be admitted in societies like India's due to certain cultural barriers. Care-giver stress can lead to mental problems and may force them to engage in potentially harmful behaviours. Potential harmful behaviour is significantly high among care-givers who are depressed, who experience high levels of anger and resentment (MacNeil et al. 2010). Many care-receivers are likely to experience neglect, abandonment or abuse at the hands of a stressed care-giver. Well defined and distinctive strategies for livelier longevity will ensure respect, dignity and worth in old age without victimization, mistreatment and abuse at the familial and societal levels.

Empirical studies on elder care indicate that in the contemporary, changing Indian family context, elder care and abuse (in any form) are intertwined. In most cases it is latent, which is invisible, unattended, unnoticed, unacknowledged and under-reported (Jamuna 2003; Khan 2004; Veedon 2001). Research evidence show that despite these changes, family ties, especially inter-generational linkages in India are still strong and the majority of elderly people still live with their families. Yet, the quality of their family life is a moot issue. Evidence indicates that since 1970, there have been a number of case-study reports on mistreatment and abuse of senior citizens residing in their homes, but studies on elder abuse in residential care facilities in India have not been publically addressed (Ramamurti et al. 2015). Senior care homes or old-age homes are comparatively new in India, and regulations for monitoring and mechanisms for reporting of mistreatment are not in place yet (Varughese and Jamuna 2014). A study undertaken in assisted living facilities in the USA reported that elder abuse and mistreatment were associated with factors such as health status of the residents, unlicensed formal care-givers, and lack of monitoring strategies in institutions. Environmental factors such as loud noise or congestion in public spaces due to equipment (e.g. walkers, wheelchairs) may contribute to resident-to-resident mistreatment (Phillips and Guifang 2011). There is research evidence that resident-to-resident abuse is widespread in US nursing homes. The most common forms are verbal abuse, some sort of physical abuse, invasion of privacy, roommate issues and inappropriate sexual behaviour (Teresi et al. 2014). Findings from the randomized control trial suggested the importance of formal, well-developed staff education programmes on resident-to-resident mistreatment.

Older individuals themselves perceive abuse as neglect, violation and deprivation of basic rights (WHO 2002). In recent decades, elder abuse has been acknowledged as a topic of concern and a number of research studies have appeared in scientific journals and special reports such as that by HelpAge India (2014).

A study by HelpAge India, revealed that 50% of elderly participants reported that they experience some form of abuse. The percentage of elders abused went up drastically from 23% in the previous year to 50% in 2014. While elder abuse in India has gone up, it is still unnoticed and under-reported by many older adults (HelpAge India 2014). However, it is an accepted fact that incidents of abuse of all types are under-reported due to reasons of social desirability, emotional dependence on the abuser (46%), economic dependence on the abuser (45%), changing cultural ethos (38%) and norms. A commonly stated reason for not reporting abuse was lack of confidence in any person or agency to deal with the problem (HelpAge India 2014). Most elders would also prefer not to acknowledge the elder abuse in public. It is also difficult to define elder abuse in the absence of any federal standards to detect and treat abuse and neglect.

Research in the past has focused on child abuse and spousal abuse, but in recent decades the focus has been on elder abuse in view of its increasing prevalence. Studies on elder abuse with methodological variations do not share the same terminology. However, for certain, there are common threads that give us an idea of the exact nature of abuse and neglect. The term 'abuse' has been defined in many ways. The methods used in the studies vary and so do the definitions. For example, absence of personal care and supervision in one study was termed as 'physical abuse' (Block and Sinnott 1979) and in another as 'active neglect' (Douglas 1983). According to recent research reports, in contrast to 50 years ago, about 50% of older adults experience some form of abuse.

Depending on the environmental or situational factors, abuse or mistreatment is categorized under three major headings. The first category is domestic mistreatment that occurs in their home setting from a family member or from significant others. The second category of mistreatment occurs when an older adult has a contractual arrangement in formal settings and suffers abuse from a care-giver (e.g. long-term care facilities, assisted-living facilities, hospitals). Self-neglect is the third category of mistreatment, when a mentally competent senior is engaged in a behaviour that threatens his or her own safety such as refusal to eat, rejection of self-care or medical help etc. Abuse is an act of omission or commission that may harm an individual.

Although the literature categorizes types of abuse or mistreatment under various headings, it is important to know what constitutes abuse within the context of one's culture. For instance, in some cultural practices, forms of name-calling may not be considered abusive, and respecting someone's privacy entails not interfering too care much in their life. Whereas in some other cultures, it is expected that elders of the home be addressed respectfully and that they become involved in all matters concerning other members in the family. As today's society is getting more youth-oriented, it does not centre on older people and their values. In families where inter-generational exposure and sharing are not common, the presence of elders can be uncomfortable for members of younger generations. However, there are no two opinions to state that elder abuse and neglect exist in the Indian context also due to various factors affecting the society and the culture. Abuse can be physical, verbal, psychological or emotional, sexual and involve financial

Table 5.1 Types of abuse

Physical abuse	Verbal abuse	Psychological or emotional abuse	Financial abuse	Sexual abuse
• Shaking • Tying down (use of restraints) • Hitting • Threatening with objects • Pushing • Biting • Burning • Using excessive force	• Name calling • Yelling • Screaming • Verbal outbursts	• Threats • Insults • Harassment • Harsh comments • Ignoring • Silent treatment (not talking)	• Scams • Misuse of money • Transfer of money or occupying property without the consent of the owner	• Unwanted contacts with genitals, anus, mouth • Taking photograph without consent • Coerced nudity • Rape

mistreatment or exploitation (Bhargava and Kumar 2014). A cursory glance of Table 5.1 explains various types of abuse.

Neglect and abandonment are other forms of abuse prevalent against seniors. Neglect occurs by withholding food, medication and other personal care from an individual. Abandonment is a form of neglect whereby someone who has agreed to take care of the older adult (or is legally responsible for so doing) fails in his or her responsibility. Preventing an individual from making independent decisions when he or she is capable of doing so, violating someone's privacy and restricting freedom to worship etc. are forms of neglect as well. Most cases of self-neglect occur in hospitals and long-term care settings. Extreme cases of self-neglect are reported among individuals who live alone, devoid of family and community support. Lack of formal and informal supportive services, the probability of chronic medical conditions and cognitive impairment are the commonly cited reasons for self-neglect in individuals who are living alone (Rathbone-McCan 2014).

Elder care is a delicate, humane and dyadic activity. Elder abuse is the negation of this noble activity (Jamuna 2004, 2006b) and any deliberate physical or psychological harm or economic exploitation is an act of elder abuse. The contributory factors of elder abuse suggest that regardless of the cause, the person who is abused in a formal or informal setting is the older person, the care-receiver. The dominant person, the care-giver has the opportunity to exploit the other person, physically, socially and financially. This negative interaction is aggravated by an ageistic perspective exacerbated by the perpetrator's lack of sentiment and sympathy (Jamuna 2004).

Elder abuse always occurs in a context. It is appropriate to go into its anatomy in terms of predisposing and precipitating factors that modulate the phenomena of elder abuse (Fig. 5.1). A question of omissive or commissive behaviour in the interactional context, predisposing, precipitating factors in the causation of abuse are crucial in understanding the abuse in a care context. The predisposing factors could be financial viability, disability and dependency level, gender, marital status,

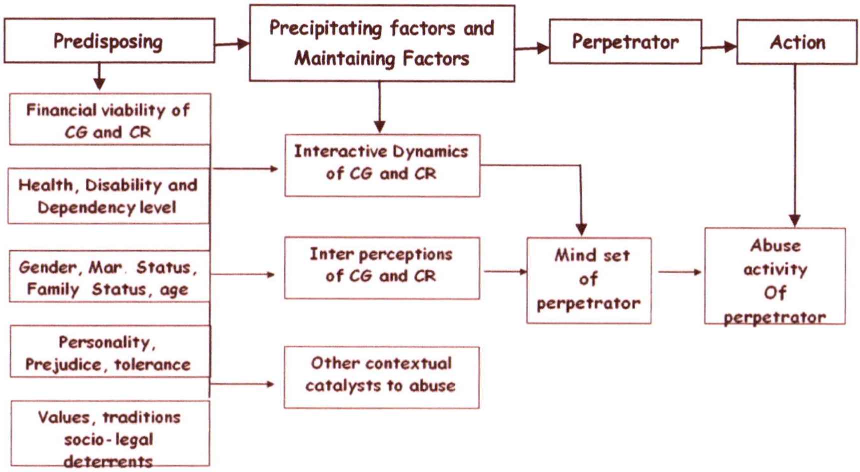

Fig. 5.1 Conceptual model of the anatomy of elder abuse in the Indian context. *Source* Jamuna (2006a)

age, personality, values, traditions, cultural factors, and socio-legal deterrents influencing the mind-set of the perpetrator. Financial viability of both care-giver and care-receiver are important factors; for instance, a care-receiver who is economically viable can afford to exchange money for care services. In this case, the economic burden will not fall on the care-giver. Similarly, the care-giver who is financially viable can have alternative provision to cater to the care needs of the elderly. The second factor that was often identified is health and disability. Obviously, the greater the dependency on the care-giver, the greater the vulnerability to abuse.

It is obvious that age per se is an important factor in elder abuse. With advancing age, fading health and frailty, consequent dependency increases the vulnerability to abuse. Analysis of profiles of abusers and victims suggest that certain idiosyncratic personality characteristics (such as aggression, dominance, exploitative nature, rigidity and sensitivity) were found to be the explanatory factors in elder abuse (Jamuna 2006b). Research suggest that abuse happens commonly to demented elders and is often perpetrated by individuals with mental diseases (Rosen 2014). Ageistic practice, prejudice and stereotyping views influence the judgement of the abuser. The transformation from collectivistic to individualistic value orientation is yet another determining factor in elder abuse. Empirical studies that the typical victim of abuse is a woman, over the age of 65 years, physically dependent for care, financially dependent, having income sources, bedridden, with behavioural problems and cognitive deficits (Shyalaja 2013). Most abuse happens when care-receivers remain alone or without care-giver's supervision, when there is a lack of accountability in the care-giving context, and when the care-receiver is unable to respond. Research reports in India also suggest that elder abuse and neglect cases vary across the socio-economic status and gender of the victims (Shyalaja 2013). However, it is reported that the likelihood of mistreatment decreases with affluence, occupation, wealth and good health (Gill 2014).

Since care giving is a delicate dyadic interaction, the factors generated during the interaction in may become precipitating and maintaining factors (for example, familial crises, power equations) that lead to emotional crises and interactional problems in the caring context. As the care-giving interaction precedes several irritating behaviours, verbal and non-verbal pop-up messages contribute to the abuse. In an Indian setting, prejudiced inter-perceptions and power equations in the relationships maintain certain activities of abuse. All these factors together determine the mind-set of the perpetrator. The state of mind converts or transforms the idea in the form of abuse, which ultimately results in different forms of abusive practices. It is difficult to measure the actual prevalence of elder abuse due to, at least in part, lack of appropriate screening instruments, and consequently under-reporting of abuse by care-receivers and care-givers (Jamuna 2003, 2004; Varughese and Jamuna 2014).

Usually, abuse happens close to home and is perpetrated by close relatives such as son or daughter-in-law. Women, often daughters, are frequently perpetrators and mothers are often victims. A qualitative research project, using ground theory design, studied the conflicted daughters in conflicted relationship with their ageing mothers. It found that 'aggression' was the main mode of interaction that often leads to an abusive relationship. The aggression which continued from childhood was the root cause of hostility, disaffection and mistrust resulting in an abusive relationship (Pickering et al. 2014). A popular survey which appeared in the Press Trust of India (29 September 2012) stated that about 31% of elders, aged 60 and above, faced disrespect, neglect and verbal abuse from their family members. The primary abuser was the son (56%), followed by the daughter-in-law (23%). In the survey, 30% of the older women compared to 26% of the older men identified the son as the primary abuser, 15% of older women compared to 8% of older men identified the daughter-in-law as the primary abuser. More than 50% of those abused had faced this situation very frequently for more than 5 years, 33% for up to 3 years and less than 1% for more than 6 years. The most common form of abuse was disrespect followed by neglect and verbal abuse. About 55% of the abused elders did not report to anyone, in an effort to maintain the family honour, the study said. It also highlighted that 20% of the abused elders were graduates, 19% had education up to primary level and 16% were illiterate.

The study conducted by HelpAge India (2013), in 20 cities with 5600 respondents, found that about 75% of the abused elders lived with their family and 69% were the owners of the house in which they were living. Of the 20 cities, the largest number of elder abuse cases reported were from Madhya Pradesh's capital, Bhopal, where 77% of respondents suffered abuse, followed by Guwahati with 60.55%, Uttar Pradesh (52%), Gujarat (42.97%), Andhra Pradesh (42.86%), Karnataka (37.14%), Jammu and Kashmir (33.5%) and Chandigarh (32.71%). Out of total respondents, 51% were women. Some 59% were in the age group of 60–69 years; 28% in the age group of 70–79 years and 8% in the age group of 80 years and above. In the age group of 60–69, there were more women than men, and in the other two categories there were more men than women. The study found that the situation was almost the same across all socio-economic categories.

A Different Category of Abuse

A 2013 study by Priyamvadha at the Department of Criminology, Madras University reported the inhuman and illegal practice of 'Thailai Koothal' in Virudh Nagar, Theni district, Tamil Nadu Priyamvadha (2013). This is a practice of gerontocide or involuntary euthanasia whereby the elderly person who is bedridden is given a ceremonial oil bath followed by ingestion of tender coconut water to induce pneumonia and to accelerate death. Interestingly, about 30% ($N = 1200$) of respondents acknowledged that gerontocide was being practised as a ritual killing of the elderly in the state, while 22% acknowledged that death-hastening ceremonies were being performed to terminate the life of the elderly. The respondents quoted a number of reasons for the illegal practice, including concern for their parents, irreversible physical and mental condition of the elderly and low economic status. They expressed that they did not want their parents to suffer during old age (49%), claimed there was no scope to regain quality of life because of their physical or mental health problems (34%) and blamed poor economic status (23%). Other unpleasant findings in these cases included 36 different methods for killing the elderly. The decision to snuff out the life was mostly taken by the immediate family, though others also played a part at times. This type of gerontocide is an act of abuse and neglect, as described by WHO (2002) and attested by the elderly themselves. The possible indicators of abuse are shown in Table 5.2.

Assessment and Screening of Possible Abuse

Whether it be a physical, emotional, sexual or financial type of abuse, it will have serious repercussions in the life of an individual. Studies report cognitive deficits, depression, anxiety, fear, fatigue and somatic complaints as the most predominant physical and mental symptoms of abuse. Physicians, nurses, caseworkers and other direct care-givers should be proficient in identifying symptoms of all types of abuse and neglect. There are multiple screening and assessment tools available but there is no single, well-accepted tool or method of assessment for elder abuse. In addition, health-care professionals need thorough training to administer these tools. However, health-care professionals such as physicians and nurses who are in direct contact with patients in hospital emergency rooms and clinics should be able to identify the potential red flags. When a comprehensive screening for elder abuse or mistreatment is not possible in emergencies, a single-question screen (see below) or three-question screen (see below) can be used to identify the triggers (see also Table 5.3).

Table 5.2 Possible indicators of abuse and neglect

Physical abuse	Emotional abuse	Financial abuse	Neglect	Sexual
• Unexplainable marks and bruises on body (at various stages of healing), open wounds and excoriations • Multiple fractures (old or new), sprains, dislocations • Contractures • Burns in the shape of an object • Injury to internal organs • Internal bleeding • Dehydration • Malnutrition • Unexplainable weight loss • Faecal impaction • Poor hygiene	• Fear—possibly in the presence of abuser • Agitation • Resisting care • Resistance to answering questions • Confusion • Depression • Social isolation • Insomnia • Irregular patterns of sleeping • Significant weight loss or gain	• Misuse of money • Inability to remember account or property details • Transfer of money to a different account • Changes in written will • Living well below means • Discomfort when discussing finances	• Poor appearance • Poor hygiene • Malnutrition • Poor dental hygiene • Body odour • Bedsore • Inadequate or inappropriate clothing	• Bruises on or around genital area/breasts • Unexplained vaginal or anal bleeding • Pain on walking/sitting • Unexplained sexually transmitted diseases or genital infections

Single question screen: Are you being abused or neglected?

Three-question screen: (i) Do you feel safe where you live? (ii) Who prepares your meals? (iii) Who takes care of your chequebook?

Suspected elderly victims of mistreatment should undergo a comprehensive geriatric assessment in addition to history and physical examination. A comprehensive examination including evaluation of mental and emotional status, mood and behaviour along with functional capabilities and limitations, general appearance etc. is critical. Health-care professionals must determine the person's level of cognitive status and the ability to answer questions. Assistive devices such as eyeglasses, hearing aids, and dentures are to be provided to enhance communication. Assessment of abuse is difficult if the victim has impairments in cognition and communication. It is very important to assess the signs and symptoms of abuse and neglect and ascertain whether the individual is in any immediate danger. Comprehensive assessment of elder abuse needs a multi-disciplinary team approach. Victims seeking medical care for the symptoms described above should undergo a thorough history taking and physical examination. A head-to-toe inspection should include visual examination of the body and skin for any evidence of physical abuse. Further diagnostic tests such as radiological examination should follow if there is suspicion of fractures or internal injuries; neurologic examination to assess cognitive status; genito-urinary and rectal examination for signs of sexual

Table 5.3 Additional screening questions suggested by the American Medical Association

Questions for the abused	Questions for the possible abuser
Is someone hurting you physically or verbally? Have you been threatened or called names?	Tell me about the person you give care to and how much care that person requires
Do you feel you are being cared for properly?	What tasks can the person perform by themselves and what are the limitations?
Have you ever been pressured to do something against your will?	What expectations does the person have for you? Can you accomplish what he/she expects?
Do you have any concerns about your finances? Has your money or property ever been used by someone else? Has any of your money or property been taken or signed over to someone else?	Do you have any responsibilities outside the home?
Does your care-giver rely on you financially?	Elder person's care can be quite frustrating. Has the person ever frustrated you?
Are you aware of any behavioural violence, alcoholism, drug abuse, psychiatric disorders by your care-giver?	Have you ever raised your voice or threatened the patient? What kind of support would you need in order to help take care of the person?

Source Carmen and Lachs (2014)

abuse (Tabloski 2013). Laboratory tests can confirm the presence of dehydration, malnutrition and anaemia. It is important to explore a person's living arrangements and ascertain how long he or she has been living there, who owns the home and who pays the bills. Questions regarding financial status can provide crucial information regarding potential financial abuse. Social support for the victims in the community should be determined, as the abuse victims are usually isolated. The mental-health practitioner is a crucial member of the interdisciplinary team in providing support services for elders that promote self-efficacy and belief in their own ability to look after themselves. Rathbone-McCan (2014) suggests practitioners focus on six dimensions of self: (a) sense of self; (b) sense of will (c) awareness of capacity (d) awareness of improved capacity (e) awareness of existing environmental supports; and (f) awareness of supplementary assistance.

Intervention Strategies

Medical Intervention

Documentation and Reporting

Due to the absence of reliable national data, the prevalence and incidence of elder abuse in India is not very clear. Incidents are mostly hidden, and/or under-reported due to the sensitive nature of the topic (Shankardass 2013). Therefore, it is of great importance that health-care providers document the chief complaints of the abused person in his or her own words (subjective data). Documentation should include a complete medical and social history of the patient including injuries present (number, type, location), all laboratory or radiological and imaging studies conducted, and photograph of wounds or bruises on the skin. If police are notified, then the name of the officer and the incident number should also be recorded. Each abuse case must be reported to the legal system: for example, Adult Protective Legislations or Adult Protective Services depending on the mechanism available. It is worth noting that in developed countries like the USA, health-care providers are mandated to report suspected abuse or mistreatment or else they will be found negligent in their responsibility.

Social Service and Legal Interventions

Health-care providers should ensure that the medical and social needs of the individual are met before they are discharged from the institution. Interdisciplinary team members (especially the social worker) must ensure that the home environment is safe for them to return to. A safety plan should be designed with the help of legal and protective services, ensuring the home environment is safe and there is care available for physical and mentally impaired individuals. The social worker should also ensure that the interdisciplinary team continues to follow up the elder in terms of his or her physical and mental functioning, availability of resources and social support.

In India, intervention can be carried out by modulating (a) predisposing and (b) precipitating factors of vulnerable groups such as: older women; destitute, homeless, frail, disabled and widowed elderly people. Special privileges in terms of housing, clothing, medicine, food and assistive devices should be provided to these groups. Home-health agencies can provide routine home assessments by community social workers and/or nurses. Provision of legal protection to victims and social programmes directed at changing the mind-set of the perpetrators through personal and family counselling supports should be prioritized. At a community level, it can be accomplished—through effective modelling, raising awareness through print and electronic media. Interventions may be aimed for the short term, or long-term custom-built and contextualized to meet the specific demands of the situation.

Financial viability of the elderly through a Universal Old Age Pension Scheme and health insurance to cover the cost of sickness and disability are necessary. In the absence of health insurance, the government should reimburse the expenses incurred by the sick older persons. Improving health and prevention of disease and disability are necessary through health-promotion activities. Income-generation activities (part-time jobs, skill-related money-generating activities, long-term saving plans (e.g. LIC—Varishta Nagarika Saving scheme) are to be encouraged.

Reinforcement of legal and legislative actions concerning elder abuse prevention is vital. Most of the time, victims are not aware of protective laws and whom to turn to for help. The majority of the victims fear that complaint may bring a bad name to the family and that they may even be further isolated and punished for their actions. Identifying potential risk factors and removing them from the environment is an essential step. Sometimes abusers may not be aware that they are abusing the victims. Due to personal strain, the care-takers may mistreat the care-receiver knowingly or unknowingly. Care-givers should be able to have open discussions of their problems and concerns with a professional. Educational awareness programmes for care-givers and family members on elder abuse should be held in communities. More professionals should take the initiative in making support resources available for care-givers and care-receivers. Mental-health services should be available for both abusers and victims to determine the need for further actions. Respite care services and Adult Protective Services (APS) should be available for temporary placements of the victims. Abuse and violence prevention programmes should be a part of the curriculum in schools, colleges and universities. To prevent elder abuse, there is a need for nationwide programmes in schools and colleges for sensitizing children and young adults about the right to dignity in ageing.

The government of India is ensuring the safety of citizens through the National Policy on Older Persons (1999), National Policy on Senior Citizens (2011) and Maintenance and Welfare of Parents and Senior Citizens Act (2007). The primary aim of the Act is to ensure safety and promote livelier longevity. Sadly, only 14% of the victims surveyed by HelpAge India (2014) are aware of the Act. Building a political and social lobby in favour of elder care and against elder abuse is needed. Developing an efficient, legal-reporting system, ensuring confidentiality of the reports, assuring safety and comfort of victims, reinforcing legal consequences for perpetrators, ongoing monitoring and evaluation of interventions are important strategies in preventing elder abuse.

Over recent decades, review of the research in the USA has identified three key domains that need to be addressed to reduce the risk of elder mistreatment. These domains are: (a) research (b) services; (c) policy (Pillemer et al. 2015). Even in developed countries, there is a paucity of research and a gap in knowledge about identifying and preventing elder mistreatment and introducing evidence-based interventions. More funds need to be devoted to research in this area and the research knowledge need to be translated to effective prevention strategies. Knowledge about the link between neuro-cognitive ageing and elder mistreatment risk can provide new insights about prevention strategies for elder mistreatment.

India is witnessing a major emergence of institutions for seniors such as retirement centres, independent-living or residential homes and assisted-living facilities. Elder mistreatment is common in institutional settings as well. There is a need for policies for monitoring, and mandatory reporting of abuse or mistreatment is an urgent need in the institutional settings. Educating staff about elder mistreatment and strategies for reducing resident-to-resident aggression, improved staffing for caring for the functionally and cognitively impaired residents can decrease the risk for elder mistreatment. Elder-mistreatment training should be targeted at more people in the community such as postal workers, domestic workers, family care-givers, bank employees and community workers.

References

Bhargava, A., and Y. Kumar. 2014. Elder abuse in Indian families: Problems and preventive actions. *International Journal of Scientific and Research Publication* 4 (10): 3498–3506.

Block, M.R., and J.D. Sinnott. 1979. *The battered elder syndrome: An exploratory study*. College Park, MD: University of Maryland Center on Aging.

Carmen, T., and M.S. Lachs. 2014. Detecting, assessing, and responding to elder mistreatment. In *Current diagnosis & treatment: Geriatrics*, 2nd ed, ed. E.A. Williams, A. Chang, et al., 517–523. New York: LANGE: McGraw Gill Education.

Douglass, R.L. 1983. Domestive neglect and abuse of the elderly: Implications for research and service. *Family Relations* 32: 395–402.

Gill, R. 2014. Elderly and caregivers. Some narratives of abuse. *HelpAge India Research and Development Journal, Special Edition on Elder Abuse and Dignity* 20 (2): 41–47.

HelpAge India. 2013. Identification, intervention and prevention of elder abuse: A manual for professional and volunteers. *HelpAge India Research and Development Journal* 6–20.

HelpAge India. 2014. Main findings, report on elder abuse in India. Retrieved from: http://www.who.int/ageing/projects/elder_abuse/alc_ea_ind.pdf.

Jamuna, D. 1995. Elder care: The developing Indian scenario. In *Challenges before the elderly: An Indian scenario*, ed. S. Vijayakumar, 257–272. MD Pubs.: New Delhi.

Jamuna, D. 2003a. Issues of elder care and elder abuse in Indian context. In *An aging India: Perspectives, prospects and policies*, ed. Phoebe S. Liebig. New York: The Haworth Pres, Inc.

Jamuna, D. 2003b. The use and abuse of care: What the caregivers and care-receivers say? In *Elder care: Emerging trends, challenges and perspectives*, ed. K. Ganagadharan. Hyderabad: Heritage Hospitals and Indian Institute of Health and Family Welfare.

Jamuna, D. 2004. The statics and dynamics of elder care in the Indian context. In *Handbook on Indian gerontology*, ed. P.V. Ramamurti, and D. Jamuna, 208–242. Delhi: Serial Publications.

Jamuna, D. 2006a. *An advocacy model for the prevention of elder abuse*. Centre for Research on Ageing, S.V. University, Tirupati.

Jamuna, D. 2006b. Graying of population—Policy dilemmas. In *Globalization: Contextualizing social science research*, ed. I. Ramabramham, and M. Hariharan, 51–61. New Delhi: Allied Pubs. Pvt. Ltd.

Khan, A.M. 2004. Decay in family dynamics of interaction, relation and communication as determinants of growing vulnerability amongst elderly. *Indian Journal of Gerontology* 18 (2): 173–186.

MacNeil, G., J.I. Kosberg, D.W. Durkin, W.K. Dooley, J. DeCoster, and G.M. Williamson. 2010. Caregiver mental health and potentially harmful caregiving behavior: The central role of caregiver anger. *The Gerontologist* 50 (1): 76–86. doi:10.1093/geront/gnp099.

National Policy on Older Persons. 1999. New Delhi: Ministry of Social Justice and Empowerment.

National Policy on Senior Citizens. 2011. New Delhi: Ministry of Social Justice and Empowerment.

Phillips, L.R., and G. Guifang. 2011. Mistreatment in assisted living facilities: Complaints, substantiations and risk factors. *The Gerontologist* 51 (3): 343–353.

Pickering, C.E.Z., A. Moon, H.C. Pieters, C.J. Mentes, and L.R. Phillips. 2014. Relationship management strategies for daughters in conflicted relationship with their ageing mothers. *The Journal of Advanced Nursing* 71 (3): 609–619. doi:10.1111/jan.12547.

Pillemer, K., M.-T. Connolly, R. Breckman, N. Spreng, and M.S. Lachs. 2015. Elder mistreatment: Priorities for consideration by the white house conference on aging. *The Gerontologist* 55 (2): 320–327. doi:10.1093/geront/gnu180.

Priyamvadha Mohan Singh. 2013. Paper presented at Ageing and Society. *Third interdisciplinary conference and the Aging*, University Center Chicago, Chicago, USA.

Ramamurti, P.V., S.P. Liebig, and D. Jamuna. 2015. International spotlight: Gerontology in India. *The Gerontologist* 55 (6): 894–900.

Rathbone-McCan, E. 2014. An improved approach to treating elder self-neglect: The self-care framework. *Generations Journal of the American Society on Aging* 38 (3): 80–85.

Rosen, L.A. 2014. Where mental health and elder abuse interact. *Generations: Journal of the American Society on Aging* 38 (3): 75–79.

Shankardass, M.K. 2013. Addressing elder abuse: Review of societal responses in India and selected Asian countries. *International Psychogeriatrics* 25 (8): 1229–1234. doi:10.1017/S104161021300063X.

Shylaja, L. 2013. Determinants of elder abuse in Kerala: An empirical analysis HelpAge India. *Research and Development Journal* 20 (3): 5–16.

Tabloski, P.A. 2013. *Clinical hand book for gerontological nursing*. Upper Saddle River, NJ: Pearson.

Teresi, J.A., M. Ramirez, J. Ellis, S. Silver, G. Boratgis, J. Kong, et al. 2013. A staff intervention targeting resident-to-resident elder mistreatment (R-REM) in long-term care increased staff knowledge, recognition and reporting: Results from a cluster randomized trial. *International Journal of Nursing Studies* 50 (5): 644–656. doi:10.1016/j.ijnurstu.2012.10.010.

The Maintenance and Welfare of Parents and Senior Citizens Act. 2007. New Delhi, India.

Varughese, R.A., and D. Jamuna. 2014. *Health of residents of senior care homes*. Germany: LAP LAMBERT.

Veedon, R. 2001. Elderly abuse in urban context. *Indian Journal of Social Work* 62 (3): 480–491.

World Health Organization. 2002. *Active aging: A policy frame work*. Geneva: World Health Organization.

Chapter 6
Invisible Suffering of Elderly People Among Families in Kerala: Neglect and Abuse

S. Irudaya Rajan and U.R. Arya

Introduction

The socio-economic and technological changes in the world have placed old people in misery. Not only have the younger generation shifted from a joint-family system to a nuclear one, but also they have waved goodbye to their cultural values and tradition in relation to care given to elderly parents. As part of adopting a new culture, they have been prepared to abandon their old parents who once took all kinds of trouble to selflessly feed and care for them.

Gone are the days of the belief that 'old is gold'. There was a time when an elderly member of the family was vested with all authority and decision-making powers. So important was the consent of the family's older members that not for any reason would younger ones in dare to take independent decisions without consulting them. Now everything is changing, and so are people. In their race to reap material wealth, people often treat their old parents as a burden. It is in these instances that young people try to get rid of such a burden by means like neglect, abuse and abandonment. Old people are frequently put under house arrest, with their freedom of movement restricted, too. It is often believed that a person is safe within the four walls of his/her house. But this notion has received a major blow with the cases of abused and neglected elderly people increasing day by day. Ageing of the population coupled with increased longevity are the root causes at this juncture.

Neglect is the refusal to fulfil a care-giving obligation (WHO 2002), whereas abandonment of elderly people is the desertion of them by an individual who has

S. Irudaya Rajan (✉)
Centre for Development Studies, Thiruvananthapuram, Kerala, India
e-mail: rajan@cds.ac.in; rajancds@gmail.com

U.R. Arya
Centre for Development Studies, Thiruvananthapuram, India
e-mail: aryaur1111@gmail.com

Mala Kapur Shankardass and S. Irudaya Rajan (eds.), *Abuse and Neglect of the Elderly in India*, https://doi.org/10.1007/978-981-10-6116-5_6

physical custody or otherwise has assumed responsibility for providing care for them (Gorbien and Eisenstein 2005). Elder abuse is different from both neglect and abandonment, though it may include such acts. As per WHO (2017), elder abuse can be defined as 'a single or repeated act, or lack of appropriate action, occurring within any relationship where there is an expectation of trust which causes harm or distress to an older person'. In Kerala, 3% of people fall victim to abuse after the age of 60 (UNFPA 2011).

The dwindling of the joint family, the rise of dual-career families, a possible shift in filial piety values, the increasing life expectancy with greater chances of a prolonged old age characterized by poverty, degeneration, more empty-nest years, and dependency, have all added to the seriousness of the problem and made the elderly more susceptible than ever to abusive treatment (Jamuna 2003). It is found that fewer members in the household reduces the chance of abuse while having a greater number of children increases the chance of abuse (Vegard and James 2014).

Under the Maintenance and Welfare of Parents and Senior Citizens Act, 2007, a senior citizens (including parents) who is unable to provide maintenance for himself/herself from his/her own earning or out of the property owned by him/her shall be entitled to get support from his/her children because, as per this Act, the obligation of the children to maintain their parent extends to the needs of such a parent (either father or mother or both, as the case may be) so that such a parent may lead a normal life (Ministry of Law and Justice 2007).

The share of elderly population in Kerala's total population is more than 12%, and 58.90% of households have no elderly person (Census 2011). In India, Kerala has the largest number of old-age homes, with the number increasing. Old-age homes, which used to be the last resort, are now the first choice of the new generation (Sreeraj 2016). In many cases, though old people are not willing, children forcefully dump them in old-age homes. In some other cases, old parents prefer to stay in old-age homes rather than suffer at the hands of their children. Some 0.3% of elderly people prefer staying in old-age homes to staying with family members (UNFPA 2011).

In such a scenario, this chapter tries to examine the risk factors leading to elder neglect and abuse within the family. Attempts are made to discover the relationship of the abuser with the abused elderly person. Different forms of elder abuse within the family are studied, as are the consequences of elder abuse by a family member.

Focus of the Study, Data and Methodology

There have been many newspaper reports of elderly people suffering neglect and abuse in Kerala. Seniors' living patterns are important for understanding the relationship between the abuser and the abused. Among those who live with a son, 56.4% are females and 56.2% are males. By contrast, the proportion of elderly females living alone is 4.3 and 1.2% elderly males (Rajan et al. 2017). Understanding such living arrangements is particularly crucial because abuse nowadays often comes also from those people who are supposed to take care of the elderly during their twilight years. This highlights the fact that elderly people are not safe even within the four walls of their own homes.

Unlike other topics, it is very difficult to find data on elderly neglect and abuse as these are the cases which cannot be expressed purely in numbers. Also, many people are afraid to reveal the exploitative situations they face in life. This happens especially when the culprits are their own children. Only reported cases are known. There are many unreported cases which are many times more frequent than the reported ones. Even in the case of reported instances, a single number cannot express the type of traumatic and painful experiences which old people have gone through. Thus, to capture the sufferings of elderly people due to abuse and neglect within the family, we conducted case studies in and around the district of Thiruvananthapuram. Observation and semi-structured interviews were also used here. The samples in the study include 20 old people falling in the age group of 60–90. Both males and females were covered in this study (see Table 6.1).

The study is explorative in nature. Elderly people living with their family members form one group of the sample. Seniors living in old-age homes are also included in the study to portray the real picture of the nature of miseries they faced at home because of mistreatment from family members. Snowball sampling is used in this study. Insights drawn from case studies are rich in instances of abuse, neglect and hidden sufferings of elderly people. Since many of the samples in this study are still worried about the future of their children, children-in-law and grandchildren, they do not want their names to be disclosed. Narrative style is used to comprehend the miseries of the elderly as this method is best suited to unveil the kind of oppression and exploitation suffered by them in their sunset years.

Table 6.1 Brief details of the cases studied

Characteristics	Category	Number
Age	60–69	5
	70–79	10
	80+	5
	Total	20
Gender	Male	6
	Female	14
	Total	20
Marital status	Married	8
	Widowed	12
	Total	20
Living arrangement	With children or spouse or other relatives	8
	Old-age home	12
	Total	20

Source Case studies

Findings of the Study

Risk Factors Leading to Elder Neglect and Abuse Within the Family

A risk factor is any attribute, characteristic or exposure of an individual that increases the likelihood of developing a disease or injury (WHO 2017). Risk factors make elderly people vulnerable to neglect and abuse. Failing health and aching bodies demand somebody to take care of them. Thus, it is important to devote time, effort and money to provide for full care of seniors. It is the working-age group, sandwiched between the demands of their old parents and their children, who confront challenges because of stress in multiple areas such as home, office and others. It is primarily this anger that is vented on old people in the form of neglect, abuse, and sometimes abandonment. Adding fuel to the fire, old people have a high chance of developing many diseases, making this period of life worse than any other.

Risk factors operate at multiple levels. WHO (2017) has identified risk factors functioning at four levels: individual level, community level, relationship level and socio-cultural level. Risk factors can be looked at from two different angles, one related to the functional impairments of elderly people and the other to the difficulties and stress of the care-giver in looking after elderly people in the family. Many studies have proved that the chances of disability increase as one reaches old age. Disability can be single, double or multiple, at times related to vision, sight, movement, etc. As per the findings of the Kerala Aging Survey (2013), the share of elderly people suffering from difficulties related to vision, chewing, hearing and walking are 79, 44.3, 23.6 and 34% respectively. Chances of other health problems like heart diseases, arthritis, urinary infections, etc. cannot be neglected in old age. Adding insult to injury, another headache at this point is life-style diseases such as diabetes, obesity, anaemia, etc. All these make elderly people vulnerable to problems such as abuse, neglect or even abandonment by family members at a time when they require greater care and attention. It is also found that elderly people who are bedridden are the ones who suffer most.

Below we narrate the experience shared by a 77-year-old man.

> Since he was not educated, it was impossible for him to get a well-paid permanent job. So, he worked on other people's paddy fields. When it was off season he was employed by rich people to help them take care of their estates. He never wanted his children to inherit his job. So, he sent them to the nearby school. Giving no food to his own belly, he gave his children enough food. Now, both his son and daughter-in-law are doctors. After the death of his wife, they told him to sell off his property and go with them. He did so but now they do not want him, addressing him as good for nothing. Now he is suffering from kidney failure. So, they have abandoned him. They are running after money. Though his children are financially well off, they do not have time for him. He complains about his fate.

Here, lack of adequate education and the death of his wife act as risk factors, making his children feel that nobody could question their acts. Adding to his misery, he is a kidney-patient.

Another danger to which elderly people fall prey is falls. As per the findings of the Kerala Aging Survey (2004), 33.6% of seniors have not recovered fully from the outcomes of falls. Falls among elderly people in most cases lead to disability. Locomotive disability is the worst of all because a single fall can lead to an active old person becoming bedridden, making his or her life hellish. In many cases, a fall may lead to death after months or years of suffering. Thus, these types of falls are called terminal falls. Since old people have less calcium content in their body, even a trivial fall can lead to fracture.

When conditions like disability or memory loss prevail, more care is required from the care-giver. If the care-giver should manage household responsibilities, too, this may create over-strain released in the form of abusive behaviour against the elderly person in the family. Such a situation is exacerbated when coupled with financial dependency of the elderly person.

In the early theoretical models, the level of stress of caregivers was a risk factor linking elder abuse with care of an elderly relative (Eastman 1994; Steinmetz 1988). In many cases, striking a balance between household duties and work pressure appears to be a Herculean task for many of the nuclear-family members where both husband and wife are employers. In such a scenario, though not deliberately, they vent their anger on the old people at home, without even remembering that it was these old people who strived to help them to become what they are today. At times, such abusive behaviour is momentary. But one cannot take back words that are already spoken and stones that are already thrown because one cannot take back somebody's pain which has already been suffered and tears which have already been shed.

Relationship of the Abuser with the Abused

In the context of the erosion of joint families and the simultaneous rise of nuclear families, the presence of elderly people in the family is not encouraged. In many cases reported in the study, abusive behaviour towards seniors in the family came especially from the victims' own children, children-in-law and grandchildren. The sandwiched generation tries to get rid of the elderly people from their house, deliberately forgetting the care and support they once received from those hands which are now old, weak and frail.

The experience of a 90-year-old woman is narrated below:

> She suffered a lot. Her daughter-in-law tried all her means to drive her away from home, using abusive language and never letting the old woman talk to her son, who came home only once a month. On many occasions she asked the old lady to commit suicide if she

could not die normally. She used to hit the old lady on the head. Food was not given on time. She withheld all medicines to hasten this frail old woman's death. Though this old woman had seven children (six daughters and one son), none of them was ready to take care of her. One day, her daughter-in-law drove her away from home. The old widow touched her feet telling her not to abandon her. When her son came, the daughter-in-law told him that it was the old woman who had tried to leave home, shifting all blame. It was on that visit home that her son put her in an old-age home. All the residents of this new home have had similar experiences.

Though elderly people dance to the tune of their grandchildren, nobody can predict what is going to happen in future. The story of an 86-year-old lady is narrated below:

Her grandchildren left her after locking her house, taking all the furniture with them. She was 80 by then. They pulled her out of the house and drove her away. After spending weeks on the streets, she is now in an old-age home. She suffered a lot. She was shivering under rain in the open air at night, with no roof over her and no walls to protect her. They treated her like this because she was opposed to their wish to sell her property before her death. She had four children: three sons and one daughter. Now her sons are no more.

Siblings, members of the extended family or even a spouse can also be culprits of elder abuse. In most of these cases, the reason would be financial. The role of marital status is thus important when it comes to determining whether an elderly person receives adequate care and attention because this also deals with the living arrangement of the elderly. As per the findings of the Kerala Aging Survey (2013), out of the total unmarried elderly people, 38.7% were living alone whereas in the case of widowed elderly people, 47.5% were living with their son. Some 42.9% of the married elderly were living with their spouse.

Occasionally, shocking events related to servants harming and killing elders for money also happen in Kerala. Because of both out-migration and emigration increasing in Kerala, people are now more likely to leave behind their old parents. Since women also migrate to other areas in search of work, it is the paid servants who extend a helping hand to the elderly at home in many such cases. A 69-year-old woman has lot to say about how she suffered at the cruel hands of her servant:

She was left alone in a big house by the children when they moved to Canada. They also made all the necessary arrangements for a paid servant-cum-home nurse to take care of all the needs of their mother. But the servant was driven by her greed for money which was evident from her behaviour. She, in the name of dusting and cleaning, tried many times to steal money and gold ornaments. She never bothered to care for the old woman. She never gave the old woman food and medicines on time. There were many unreasonable withdrawals from the elderly woman's bank account. When this old woman tried to question her, she reacted violently by pushing the old lady to make her fall.

Types of Elder Abuse

Elder abuse may take various forms ranging from mere verbal abuse to physical and mental torture and exploitation. In most of the cases reported in Kerala, financial matters are a factor. Since the older generation has no other option but to suffer silently, they try their level best to tolerate these kinds of behaviour from their family members.

Verbal abuse ranges from ordinary scolding to the use of rude and offensive language. Expressions like 'remember that you are old' and 'stop doing that' are certain commands which are used by young people to restrict old people from engaging in some activities which give the latter pleasure. Shouting, yelling and making insulting or disrespectful comments to elderly people are also instances of verbal abuse.

Physical abuse and exploitation include beating, punching, deliberate pushing to make elderly people fall, and use of weapons. These, in many instances, lead to disability. In extreme cases, they can even lead to death. Instances of putting elderly people in chains, accommodating them in cattle stalls or a bathroom, or under house arrest have even been reported in Kerala.

Financial abuse and exploitation of elderly people is very common in Kerala. Elderly people who live alone are vulnerable to crimes such as attack (and, in extreme cases, murder), while resisting robbery.

We narrate one old man's experience below:

> His children are away in the city, living with their family who are very keen on education of their children. He is alone in his home. He is a heart-patient too. Since there is a chance of being robbed, he is afraid. Two instances of robbery have happened at his home. On one occasion, he got injured. Everyone knows that he is alone there.

Since people are driven by financial motives, they try all means to acquire wealth. In such instances they do not care whether their acts create problems for their old parents.

A 78-year-old woman reveals her suffering:

> She and her husband were living in their home. Since their children had migrated for the convenience of work and studies of their children, these elderly parents were left behind. After her husband's death, she was taken to her son's home on one condition. He and his wife asked this old woman to transfer her house and property to their name. It was almost at the same time that her daughter also asked her for the same. She never wanted her children to fight each other. So, she told them that it is for their children that parents acquire all wealth. This was considered by them as an act of refusal on her part. From then onwards, days passed like decades for her. She came across all kinds of verbal abuse from her

children. At times, they even dared to hit her and punch her. She was not taken to a doctor when she was unwell. She understood that she appeared a burden to them. She came to recognize that they were not waiting for an opportunity to take care of her but they were in a tug of war for her assets. She was convinced that if she became unwell and bedridden no-one would take care of her. So, she decided to transfer her house to her daughter and property to her Son. She had never wanted to create problems for her children. So, she herself decided to come to an old-age home. Now, she is happy. She is not alone here. Now, she can make sure that when she dies, at least there are some people who will be moved to tears.

Another instance of financial abuse concerns a 78-year-old man:

When he and his wife came to Kerala, he was 23. They had no inherited assets. The only thing they had was a determination to work hard. They could thus reap gold on earth, turning their blood to sweat. They raised their children, treating them as the apple of their eye. They had four children. They gave their children a good education and they are now in the USA. Each has a family. This elderly couple now have money, fame and everything. But though they wish to live with their children they know that this is just a dream which cannot be made a reality. After the death of his wife, his children took away his property saying that they would take him with them. But nothing happened after the transfer of property. Now he is in a destitute home.

Many elderly people who stay with their children or grandchildren or other relatives are not free from similar sufferings. Denying the older person access to his or her own assets or home, fraud, taking money under false promises, counterfeit, theft, forced property transfers, use of false signatures to deal with financial matters, improper use of legal agreements which contain the name or signature of the elderly etc. are some of the ways elderly people are financially abused and exploited. Thus, all illegal use of an elderly person's assets without his/her permission fall under the category of financial abuse.

Though elderly people can tolerate all these, they wish to stay with their family because they do not want to be neglected. Neglect, in one instance or the other is a part of life. This is true especially for an elderly person. He/she, on many occasions, is not allowed to appear in front of guests. In many situations, even when the whole family leave the home for a leisure-trip they deliberately avoid the elderly member in the family, finding some excuses like 'there is no space in the car'. Elderly people in the family, especially those who are bedridden, should be provided with food, water, medicine and clean clothes. Timely care is essential here to support them emotionally. Family care-givers may inadvertently neglect their older relatives because of their own lack of knowledge, resources, or maturity, although this is a less frequent form of abuse (American Psychological Association 2012). In many areas, elderly people are living alone. Even if they die, it is only after 2 or 3 days that others come to know this news.

Abandonment is the most extreme and common form of elder abuse found in Kerala. There have been instances of seniors being abandoned at public places, and police registering cases against people for failing to take care of their parents (Philip 2013). This type of desertion happens without seeking the consent of the elderly people. Even their own children dare to abandon their old parents for the sake of a stress-free life. They consciously and purposefully forget that what they are now is because of their parents' sweat. Kerala has a lot to say about abandoned old people. Railway stations, bus stands, government hospitals, crowded temples, etc. are the main places where seniors are abandoned.

An 88-year-old woman reveals her experience of being abandoned"

> She came with her son, daughter-in-law and grandchild to take part in festival celebration in the temple. They abandoned her in the crowd. Pointing to a tree, they told her to go and sit under it for shade to avoid tiredness so that they could go and get her something to eat. They told her that they would come back soon. She waited for hours but they did not come to take her with them. She was wandering in search of them. She pleaded with many people to help her. Somebody asked for her son's phone number. But she had no idea about it. She heard some people murmuring that her son had intentionally avoided her. She could not resist her tears. But later, she realized the truth when nobody came in search of her. From that it can be seen that they had planned to abandon her. Now she is happy in an old-age home.

All these types of behaviour, whether intentional or unintentional, can lead to the most disastrous form of abuse, which is called emotional abuse. Unlike all other forms of abuse, this is something which has enduring effects because the trauma of all the abusive experiences mentioned above can have harmful impacts on the mind of the elderly victim rather than their body. Though the wounds created on the body disappear with time, the wounds formed in the mind will stay with elderly people till their end.

Effects of Elder Abuse

Elderly people who fall victims to abuse, neglect and abandonment not only suffer from the visible or audible form of pain but also carry a sense of mental distress with them. Some after-effects of elder abuse are quite intense and long-lasting effects. These are problems which cannot be cured with medicines. If a senior is a victim of continuous abuse, it can affect his or her mind forever and there is a strong chance that they will not be able to lead a normal life again.

Intense fear and anxiety are seen among the people who once fell prey to these kinds of experiences. Because of such experiences from their own children, who are meant to take care of them during their old age, there are high chances of developing mistrust towards others. Adding to this is the feeling of isolation. Old age is the time when a person goes back to his/her childhood days. Thus, this is the time when old people need more care from young people. Because of isolation, old people may develop a feeling of hatred towards their own life.

Also, the main cause leading to depression is the feeling of isolation. For elderly people, especially those who are bedridden and left alone in a room, devoid of any human company, their only relief may be a visit by the servant. Thus, they are forced to stay away from family and friends though they all are under the same roof. They do not have anyone to open their mind. Depression occurs as an after-effects of abuse, neglect and abandonment. The scale of depression among elderly people in Kerala is shown in the Kerala Aging Survey (2013); this shows 10.5% of seniors are vulnerable as they confront difficulties due to mental ill health because of depression.

Loss of hope is also seen among old people who are avoided by their children. Therefore, they do not wish to live any more. Thus, they live as if they are waiting for their last call. Though their children abandon them, till their last moment, they at times hope that their children at some time will realize their love for them.

A hopeless situation experienced by a 79-year-old person in an old-age home is described below:

> He does not have any complaints. He does not have any regrets. He believes that his fate is to count his days there and his destiny is to spend his last days in a destitute home though he has children and other relatives who are financially affluent. Now he has no hopes and desires. Now he sees nothing else in his dreams but his death.

All these can cause fear, mental anguish, or emotional pain or distress. Suicidal tendencies increase among elderly people mainly due to feelings of isolation and resulting depression. These kinds of feelings come especially after facing the demise of a spouse. From then onwards the person is forced to live without much company. The only relief is old memories. Suicides are seen more among elderly people who live alone (Nair 2016). They commit suicide because they are suffering from multiple aliments and are sure that their children will not turn up to help them if they become bedridden. Among the elderly who have committed suicide, 78.43 and 17.98% have done so due to health problems and family problems respectively (National Crime Records Bureau 2014).

Loneliness causes unbearable pain for elderly people. Depression and suicide are the methods adopted by the mind to deal with the problems of isolation and suffering. Seniors are left with no option because many of them who stay with their children and face multiple kinds of abuse do not wish to share their experiences of hardship and suffering with others, mainly because they do not wish to defame their children.

Conclusion

Old age is the age of misery and pain due to multiple factors like deteriorating health, a tender body, low immunity, high chances of disability, etc. Thus, members of the older generation are treated as liabilities even by own children. The case studies included here also reveal that many of the residents in old-age homes have been driven out of their houses by their own children after had all their wealth snatched away. Old parents are made to sign documents of property-transfer through compulsion, threats and fraud. Once children realize that all the property has been transferred, they try, in many cases, to abandon their old parents. In many cases, the residents of old-age homes are in the state of widowhood. The strict implementation of the Maintenance and Welfare of Parents and Senior Citizens Act 2007 is thus the need of the hour.

There are many instances of elder abuse which go unreported due to the fear of being embarrassed or to the fear of further abuse and threats. The cost of these kinds of abuses cannot be calculated, especially when it comes from victims' own children. During old age, most of the people are caged, tied to the rope of torture in the prison of sorrows. Many of them have transferred their assets and properties to their greedy children and now wander the streets. They have tearful stories to tell and each has unique experiences to share.

Another group of old people are blessed with all extravagances but are devoid of human company. Even if they have all the facilities they could wish for, living alone in old age is the worst situation. For many old people, living in an old-age home is better than living alone in the lap of a luxury.

Grandparents today are alien to children in the new generation. On the other side of the coin, old-age homes are mushrooming. It is not a bed of roses or cosy and luxurious living conditions that these old parents demand. All they need is just an ordinary life with their children and grandchildren around them. Though we have laws and policies, children have more responsibility for taking care of their parents than any law or government. We need day-care centres and relief centres for old people, but these should be the last option. Thus, it is high time to treat our old people with respect to save them from situations of misery and pain.

References

Census. 2011. Population Office of the Registrar and Census Commissioner, Government of India. http://censusindia.gov.in/. Accessed December 26, 2016.

Eastman, M. 1994. *Old Age Abuse: A New Perspective*, 2nd ed. San Diego, CA: Singular Publishing Group.

Gorbien, J Martin and Eisenstein, R Amy. 2005. Elder Abuse and Neglect : *An Overview. Clinics in Geriatric Medicine*. doi:10.1016/j.cger.2004.12.001.

Jamuna, D. 2003. Issues of elder care and elder abuse in the Indian context. *Journal of Aging & Social Policy*. doi:10.1300/J031v15n02_08.

Kerala Aging Survey. 2004. Centre for Development Studies, Thiruvananthapuram.

Kerala Aging Survey. 2013. Centre for Development Studies, Thiruvananthapuram.

Ministry of Law and Justice. 2007. The Maintenance and Welfare of Parents and Senior Citizens Act.

Nair, Unni. 2016. SwathunalkiTheruvilayi. *Suprabhatham*, October 22.

National Crime Records Bureau. 2014. Suicides in India http://ncrb.nic.in/StatPublications/ADSI/ADSI2014/chapter-2%20suicides.pdf. Accessed November 18, 2016.

Philip, Shaju. 2013. In 'ageing' Kerala, living and dying alone is a trend. *The Indian Express*, August 27.

Rajan, S.I., et al. 2017. Elderly care and living arrangement. In *Elderly care in India: Societal and state responses*, ed. S.I. Rajan, and Balagopal Gayathri, 95–116. Berlin: Springer.

Sreeraj, K. 2016. OduvilOttayakunu. *Mathrubhumi*, October 2.

Steinmetz, S.K. 1988. *Duty Bound: Elder Abuse and Family Care*. Thousand Oaks, CA: Sage.

UNFPA. 2011. Report on the status of elderly in select states of India. http://www.isec.ac.in/AgeingReport_28Nov2012_LowRes-1.pdf. Accessed November 18, 2016.

Vegard, Skirbekk, and K.S. James. 2014. Abuse against elderly in India: The role of education. *BMC Public Health*. doi:10.1186/1471-2458-14-336.

World Health Organisation. 2002. Abuse of the elderly. http://whqlibdoc.who.int/publications/2002/9241545615_eng.pdf. Accessed December 26, 2016.

World Health Organisation. 2017. http://www.who.int/topics/risk_factors/en/.modified2017.

Chapter 7
Elder Abuse in Maharashtra: Extent, Forms, and Policy and Programmatic Implications

S. Siva Raju and Nidhi Gupta

Introduction

Elder abuse (the neglect and ill-treatment of persons aged over 60), though understood as one of the offshoots of inter-personal violence, is now gaining prominence and is an important public health concern and a developmental issue. Prevalence studies concerning elder abuse have so far been conducted in few developed nations and there has been no systematic collection of statistics in the developing countries. However, crime records, journalistic reports, social welfare records and small-scale studies do provide evidence that abuse, neglect, and financial exploitation of elders are widely prevalent. Elder abuse occurs in a variety of settings and can take different forms: physical, sexual, psychological, financial, neglect and abandonment (Gupta 2016). It is important to highlight that it is both under-recognized and under-reported. Many factors contribute to elder abuse and neglect. There are numerous, generally unacknowledged ways in which gender and gender relations enter into the maltreatment and neglect of older people. Women are more commonly victims of almost every type of abuse than men, and people age 80 or older are at greatest risk of being neglected. Neglect is more commonly perpetrated by women, but men are more often responsible for all other types of abuse. Most cases of elder abuse involve family members in a care-giver role. There is a greater risk of abuse in families with a history of conflict, especially if an adult is taking care of an elderly parent who was abusive to him/her as a child. Other stresses in the care-giver's life, such as marital, financial, or work-related problems,

S. Siva Raju
Tata Institute of Social Sciences, Hyderabad Campus, Hyderabad, India
e-mail: sivaraju@tiss.edu

Nidhi Gupta (✉)
Centre for Population, Health and Development, School of Development Studies,
Tata Institute of Social Sciences, Mumbai, India
e-mail: drnidhi2014@gmail.com

Mala Kapur Shankardass and S. Irudaya Rajan (eds.), *Abuse and Neglect of the Elderly in India*, https://doi.org/10.1007/978-981-10-6116-5_7

increase the risk of abuse. Use of drugs or alcohol by the care-giver also contributes. The risk of abuse increases as the elderly person becomes more frail and dependent, requiring more intensive care and placing more stress on the care-giver. Finally, there is a heightened risk if the care-giver lacks care-giving skills (Hoban and Kearney 2000).

The growth of the elderly population provides a compelling need for interventions with individual elderly and familial problems, efforts involving both institutional and societal change. There are a number of interventions in society but overall, they were deemed inadequate given the magnitude of elder abuse. It is strongly recommended that further research be undertaken so as to enable better understanding of the problem and planning for interventions (WHO and the International Network for the Prevention of Elder Abuse [INPEA] 2001).

India: Earlier Evidence

In Indian society, the situation of older persons is imagined to be safe and secure considering the value system of filial piety, conservative culture and the still sustaining joint family and co-residence with their offspring. However, the demographic trends and changing social values ushered in during the twentieth century have given rise to challenges and concerns for the older population in developing countries including India. The size of the older population (i.e. persons age over 60) was 8% of the total population in 2011, with elderly women comprising 8.4% and elderly men 7.7% (Registrar General of India 2011). Demographic transitions such as the rise in absolute numbers of older persons in comparison to other age-groups, increasing life expectancy and the onset of widowhood among women will have economic and social implications (Saxena 2006; Visaria 2001).

Older persons are being marginalized from mainstream life (nuclear family, migration of the young to towns and cities, acceptance of small-family norm), becoming dependent (as their living and health costs are to be met for an extended time) and are increasingly seen as burdensome by the younger population (Bhat and Dhruvarajan 2001; Rajan 2004; Siva Raju 2000, 2002a, 2011). The differing values and attitudes among the young due to socio-cultural change (weakening of family bonds, modern lifestyle, increased mobility), embracing of liberal values (individualism, increased entry of women into paid employment), lowered filial obligation (young people seeking independence from older ones) have raised serious questions on who is really responsible for care of the elderly (Gore 1968). Unlike in the West, in India 75% of elderly people reside in the villages. The rural–urban differences therefore have important implications for human ageing in terms of quality of life, support systems, morbidity, and attitudes towards old age (Kumar 1997).

Attitudinal studies indicate that the younger generations do not have tolerant/favourable views either towards older persons or old age (Rao et al. 2000). Thus, in the existing scenario, families are unable to adopt newer mechanisms in order to care for the elderly in the family, and due to financial limitations the government cannot provide a safety net (Shah 1999). In the light of the increasing number of families where only one or two children are available to share the responsibility of caring for the adult parents for long periods, and the paucity of resources to meet their varied needs, the likelihood of neglect and abuse towards older persons may be greater (Latha 2008). Neglect, is more commonly perpetrated by women (being care-givers), but men are more often responsible for all other types of abuse (Hoban and Kearney 2000; Siva Raju 2013).

A review of the few studies that have focused on elder abuse in the Indian context indicates that the most likely victim of elder abuse is a widowed female of very advanced age, role-less, functionally impaired, with no financial assets, lonely and living at home with someone, usually their adult child, spouse or other relatives (American Psychological Association 2012; Chokkanathan and Lee 2005; Devi Prasad and Vijaya Lakshmi 2008; Gupta 2016; Shah et al. 1995). Studies in India (Siva Raju 2000, 2002b) indicate that more women than men complain of maltreatment in terms of both physical and verbal abuse. The health profile of the elderly victims indicates that a person suffering from physical or mental impairment and dependent on the care-givers for most daily needs is likely to be the victim of abuse/neglect. It is evident that as age increases the chances of abuse/neglect also increase (Gupta 2016). Though a large section of victims of elder abuse are less educated and have no income of their own, old people with high educational background and sufficient income are also found to be subject to abuse. However, if the older person is educated up to high school level or more, the chances of being abused or neglected are almost half those of the elderly who are illiterate or with a school education below primary level (Sebastian and Sekher 2011).

Elder abuse comprises various dimensions such as physical abuse, physical neglect, sexual abuse, verbal assault, neglect of the environment and violation of rights and material abuse. Certain frequently cited explanations for elder abuse are: its cyclical nature, inter-generational transmission of violence, dependence because of physical/mental impairments, intra-individual dynamics, stress, negative attitudes towards the elderly and social isolation, family situations, care-giver issues, and cultural aspects. It is likely that abuse may occur due to the interplay of several of these factors (American Psychological Association 2012). The review of studies on elder abuse in the Indian cultural context suggests that a host of factors operate such as perceived powerlessness, social isolation, drug or alcoholic addiction, anti-social behaviour of the abuser, frailty and degree of dependency, financial status of the elderly, lack of space in the house, temperament and perception of care-givers (Prakash 2001). Regarding the profile of the abuser mentioned by the elderly respondents, they were mostly the adult children, sons and daughters-in-law together, daughter-in-law and spouse (Gupta 2016; Siva Raju 2000; Kumar 1991). Spouses were found to be the most likely abusers. In some studies daughters, grandchildren, and other relatives were identified as perpetrators of abuse. Others included servants, home nurses and care-givers. Incidences of crime against the elderly have been increasing over the years. These crimes include inflicting pain, robbery, murder, and even sexual assault.

The prevalent patterns and forms of elder abuse include mainly psychological abuse in terms of verbal assaults, threats and fear of isolation, physical violence, and financial exploitation. A study in Andhra Pradesh by Srinivas and Vijayalakshmi (2001) found that about 80% of the respondents experienced verbal abuse, 17% some sort of physical abuse, and 37% felt that their family members neglected them. In the study by Bagga and Sakurkar (2011), 25% of women above the age of 60 self-reported being abused or neglected. More than half of them complained of emotional or psychological abuse (53%), another 11% physical violence, and 6% reported experiencing abuse, violence and economic abuse and these 'environmental disturbances' seemed to affect their mental health. It was also found that 25% of perpetrators were their own family members. The study carried out in rural Kerala in India (Sebastian and Sekher 2011) identified that neglect and verbal abuse (39%) were the most common forms of mistreatment as stated by the elderly, followed by physical abuse (13%).

A recent study (Patel 2010) based on content analysis of reports published in two leading newspapers between 2004 and 2008 shows that most of the crimes against the elderly remain unreported. Female victims outnumber male ones and more crimes against the elderly were reported from urban areas as than rural ones (78 and 22% respectively). Surprisingly, 60% of the crimes were committed indoors and most of them during the day. A recent study in the USA (National Center on Elder Abuse 2003) revealed that about 16% of all elderly persons aged over 60 were abused in their families. However, it was argued that this was only the 'tip of the iceberg' and that the actual incidence was higher.

A few Indian studies give strong indications of maltreatment of elderly people in the family context (Devi Prasad 2000; Srinivas and Vijayalakshmi 2001; Usha 1997). However, there is a need to undertake studies to elicit the incidence and prevalence of elder abuse from community (non-institutional) settings in the Indian context. In India, past research has involved small samples and weak measurement of the phenomenon, and none has explored the association between elder abuse and variables such as economic status, age, health status, living arrangements, mental well-being, and so on. Similarly, no empirical evidence is available to relate how the non-abused elderly differ from the abused on these variables. Among the few studies on elder abuse in India, the HelpAge survey conducted in 20 cities where the organization is operating Elder Helplines, by interviewing 5600 seniors in 2012 shows a high incidence of elder abuse to the extent of 31% (29.46% in Maharashtra state alone).

In a study conducted in suburban Mumbai, it was observed that elder abuse has serious implications on all the dimensions of the quality of life of elderly women (Gupta 2015). The study also highlighted that elderly women who experience abuse had poor physical and psychological health and social relations as compared to older women who had never experienced abuse. It was further noted that this phenomenon was true across economic class groups. A study by UNFPA (2012) in seven states of India showed the prevalence of elder abuse is 11.4% with wide differentials across states, ranging from 1.8% in Tamil Nadu to as high as 35% in Maharashtra.

Elder Abuse in Maharashtra

Given the high prevalence of elder abuse reported in the state of Maharashtra (UNFPA 2012), the present chapter explored the phenomenon of elder abuse here to prepare a profile of elderly people who experience abuse, its forms, extent, and the role of perpetrators. Based on the study findings, a few policy and programmatic implications that will help to minimize abuse of elderly people have been made.

The study draws data for Maharashtra from the survey 'Building a,-Base on Population Ageing in India' by UNFPA (2012) conducted in the seven selected states of India. The sample survey covered a total of 1435 elderly men and women from 1198 households across the state of Maharashtra with almost equal numbers of households selected from rural (608) and urban (590) areas. To understand the profile of seniors experiencing abuse and the relation of abuse to the victims' background characteristics, bi-variate analysis was conducted by using various socio-demographic, economic, and health-related variables. Experience of abuse was considered as the dependent variable and variables like age, marital status, caste, education, work participation, land ownership, household size, and participation in social activities, health, disability, and activities of daily living (ADL) were considered as independent variables. Chi-square was used to elicit the relationship between the dependent variable and the independent variables. A few case studies were collected to substantiate various types of elder abuse and its implications for the quality of life of the older persons.

Findings

There are over 11 million elderly people in Maharashtra, who comprise about 10% of the total population of the state (Census 2011). The elderly population (60 years and above) in Maharashtra has increased by 31% and the number of the oldest old (i.e. 80 years and above) grew by 52% between 2001 and 2011. By 2026, the elderly population in the state is expected to grow to over 17 million, which would comprise 13% of the total population. Some 63% of all elderly people live in rural areas. The sex ratio among seniors in Maharashtra is 1114 women per 1000 men, which is higher than the sex ratio of the elderly population in India at aggregate level (1033 women per 1000 men). There were 5.8 million elderly women and 5.2 million elderly men in Maharashtra recorded by the 2011 Census, clearly reflecting feminization of ageing, which is progressing at a faster pace in this state than at the national level.

Profile of the Elderly

The average family size in households with at least one elderly person was 5.5, compared to 4.5 in households without an elderly person. Some 20% of the households belong to the Scheduled Castes (SC) (as determined by the caste status of the head of the household), 8% belong to the Scheduled Tribes (ST), 35% are from other backward classes (OBC), and the remainder (37%) belong to other castes.

Prevalence of Elder Abuse

More than one-third (35%) of seniors experienced abuse in Maharashtra, with a significantly higher prevalence of elder abuse reported by elder women (38.2%) than by elder men (31.5%). The prevalence of elder abuse among seniors living in urban areas (28.4%) was lesser than those living in rural ones (38.2%), and the rural–urban differences were found to be statistically significant (Table 7.1). Older women living in rural areas showed the highest prevalence of abuse, with 42.2% reporting experience of abuse, while older men living in urban areas reporting its prevalence least (26.3%).

Types of Elder Abuse: Elderly people reported experience of multiple forms of abuse. Of those who experienced abuse, the majority (91.2%) experienced verbal abuse, 69.9% reported experience of 'disrespect'. Some 64.4% experienced economic abuse and more than half of those who experienced any abuse reported experiencing 'neglect' and 'physical abuse' (55.7 and 51.4% respectively). The prevalence of all forms of elder abuse was observed to be higher in rural areas than in urban ones, and women were more likely to experience abuse than men (see Table 7.2).

Perpetrators of Abuse: On exploring the main perpetrators of abuse, it was observed that 30% of seniors reported abuse by neighbours, followed by sons (22.2%), relatives (19.9%) and daughters-in-law (16.4%) (Table 7.3).

Table 7.1 Percentage of older persons experiencing abuse according to place of residence and sex (Maharashtra, 2011)

	Rural			Urban			Total		
	Men	Women	Total	Men	Women	Total	Men	Women	Total
Experienced abuse	33.9 (114)	42.2 (152)	38.2 (266)	26.3 (80)	30.2 (102)	28.4 (182)	31.5 (194)	38.2 (245)	35.0 (448)
Total elderly (*N*)	366	395	761	315	359	674	681	754	1435
Chi square abuse* gender	7.064**			0.890			7.061**		
Chi square abuse* place of residence	13.441***								

Note The table has been prepared by authors after analysis of data from the UNFPA survey (2011)
*Statistical significance
**$p < 0.01$
***$p < 0.001$

Table 7.2 Percentage of elderly people who experienced abuse by type of abuse, place of residence, and gender

Type of abuse	Place of residence		Gender		Total
	Rural	Urban	Male	Female	
Physical abuse	74.4	25.6	45.3	54.7	51.4
Verbal abuse	73.2	26.8	42.3	57.7	91.2
Economic abuse	76.0	24.0	42.6	57.4	64.4
Showing disrespect	74.9	25.1	41.9	58.1	69.9
Neglect	75.1	24.9	37.9	62.1	55.7
Other abuse	59.4	40.6	39.4	60.6	4.9

Note the table has been prepared by authors after analysis of data from the UNFPA survey (2011)

Table 7.3 Perpetrators of elder abuse

Perpetrator of elder abuse	Percentage
Spouse	1.6
Son	22.2
Daughter	2
Daughter-in-law	16.4
Grandchildren	5.7
Relatives	19.9
Neighbours	28.9
Other	3.3

Note the table has been prepared by authors after analysis of data from the UNFPA survey (2011)

Correlates of Abuse

The socio-demographic, economic, and health-related indicators were studied in order to identify the vulnerable sections of the elderly population and the relationship between their background and experience of abuse (Table 7.4).

Age

On exploring the association between age and elder abuse, it was observed that those in the oldest old category were significantly more likely to experience abuse than the young old. About one-third in the young old category, 33.4% reported elder abuse, and in the oldest category 31.2% did so. Of the oldest old, 56.9% had experienced abuse. Disaggregation of this data by place of residence and gender

Table 7.4 Percentage of elderly people who have experienced abuse by place of residence, sex, and background characteristics, Maharashtra, 2011

	Rural			Urban			Total		
	Male (N = 114)	Female (N = 152)	Total (N = 266)	Male (N = 80)	Female (N = 102)	Total (N = 182)	Male (N = 194)	Female (N = 254)	Total (N = 448)
*Age****									
60–69 years (young old)	33.8	42.8	38.6	23.4	24.6	24.0	30.2	36.1	33.4
70–79 years (old)	28.2	31.5	29.8	25.0	44.0	34.7	27.4	35.1	31.2
>80 years (oldest old)	50.0	68.2	58.7	60.0	50.0	54.1	52.4	62.1	56.9
Total	33.9	42.2	38.2	26.3	30.6	28.6	31.5	38.2	35.0
Chi-square	7.424*	18.098***	24.027***	9.517**	11.241***	16.959***	14.456***	17.635***	30.824***
*Marital status****									
Currently married	34.6	35.0	34.7	25.7	23.4	24.9	31.8	31.3	31.6
Widowed	31.1	45.5	42.9	29.0	34.8	33.9	30.4	41.5	39.5
Other	0.0	78.6	64.7	0.0	33.3	33.3	0.0	70.6	60.0
Total	33.9	42.1	38.1	26.2	30.6	28.6	31.5	38.1	34.9
Chi-square	1.825	13.170***	11.376***	0.154	3.622	4.446	1.453	15.504***	14.673***
*Caste****									
Forward caste	39.7	44.7	42.3	26.9	37.8	32.7	35.8	42.5	39.3
SC/ST	34.5	50.7	43.2	22.9	27.7	25.7	31.1	43.3	37.8
OBC	26.5	31.1	28.7	21.7	20.5	21.1	24.9	27.0	26.0
Total	33.6	41.9	37.9	23.9	28.8	26.5	30.6	37.5	34.2
Chi-square	6.732	12.818***	17.705***	0.640	6.564	5.995	7.101	17.584***	23.379***
*Education****									
Illiterate	38.6	43.0	41.5	37.3	35.4	35.9	38.4	40.7	40.0

(continued)

Table 7.4 (continued)

	Rural			Urban			Total		
	Male (N = 114)	Female (N = 152)	Total (N = 266)	Male (N = 80)	Female (N = 102)	Total (N = 182)	Male (N = 194)	Female (N = 254)	Total (N = 448)
1–4 years	32.7	49.2	38.6	22.2	22.5	22.4	30.2	38.8	33.7
5–7 years	40.7	25.0	35.2	25.6	26.7	26.0	34.3	27.0	31.5
8+ years	20.0	33.3	22.2	21.8	15.4	20.5	20.8	22.7	21.2
Total	33.9	42.3	38.2	26.3	30.3	28.5	31.4	38.3	35.0
Chi-square	11.119*	5.834	14.753***	4.375	6.056	10.906	15.621***	9.306	28.401***
Work									
Never worked	50.0	43.8	43.9	0.0	30.6	30.6	50.0	37.2	37.3
Worked earlier	31.9	42.1	37.2	25.4	25.0	25.2	29.5	37	33.1
Currently working	35.4	41.3	37.6	27.5	39.6	31.9	33.1	40.9	36.0
Total	33.9	42.2	38.2	26.3	30.5	28.5	31.4	38.1	34.9
Chi-square	0.865	0.164	1.786	0.118	3.396	2.228	1.394	0.9	1.805
Number of household member									
One member	34.1	34.2	34.2	35.7	38.9	37.5	32.7	35.7	34.0
Two member	32.9	36.8	35.1	24.1	20.0	21.9	29.6	31.8	30.8
More than 3 member	34.4	44.6	39.6	26.0	31.5	29.0	31.6	39.8	35.9
Total	34.1	42.3	38.3	26.4	30.5	28.6	31.4	38.1	34.9
Chi-square	0.962	2.959	1.832	0.715	2.52	2.687	0.213	3.028	2.231
Children alive									
Less than or equal to 2	33.8	44.0	39.2	20.5	21.4	21.0	29.4	35.4	32.7
3–4 children alive	33.2	42.2	37.4	27.1	34.8	30.8	31.5	39.9	35.5

(continued)

Table 7.4 (continued)

	Rural			Urban			Total		
	Male (N = 114)	Female (N = 152)	Total (N = 266)	Male (N = 80)	Female (N = 102)	Total (N = 182)	Male (N = 194)	Female (N = 254)	Total (N = 448)
5 + children alive	39.4	43.0	41.3	31.8	39.0	35.9	37.4	41.6	39.7
Total	34.6	43.0	38.9	25.8	30.3	28.2	32.0	38.6	35.4
Chi-square	1.176	0.127	0.865	1.966	6.873*	8.128*	2.582	1.957	3.975
Ownership of land									
No land	33.1	47.2	42.3	25.1	31.4	28.6	29.0	40.3	35.9
Yes	34.4	34.0	34.3	32.0	15.8	25.0	34.2	32.4	33.5
Total	34.0	42.2	38.2	25.9	30.2	28.2	31.5	38.1	35.0
Chi-square	0.08	8.314***	6.621**	0.543	2.022	0.251	2.165	4.029	0.808
*Wealth quintile****									
Poorest	48.7	54.7	52.1	21.1	60.0	44.9	45.6	55.6	51.4
Poorer	33.1	35.1	34.1	32.4	40.0	36.9	33.0	35.9	34.5
Middle	20.5	25.4	22.5	31.0	19.7	25.0	24.8	22.4	23.7
Richer	20.8	45.7	33.0	23.2	23.6	23.4	22.3	31.6	27.3
Richest	13.3	14.3	13.8	22.0	26.3	23.9	20.0	23.1	21.4
Total	34.0	42.4	38.3	26.3	30.5	28.5	31.6	38.2	35.1
Chi-square	27.768***	27.971***	554.531***	2.337	20.016***	12.650**	26,579***	50.727***	77.730***
*General health****									
Poor	32.8	47.9	41.0	29.5	36.0	33.2	31.7	43.8	38.3
Moderate	44.1	49.0	46.6	24.6	31.7	28.8	38.3	43.2	40.9
Good	25.2	25.4	25.3	22.5	18.0	20.5	24.4	23.0	23.8
Total	33.8	42.3	38.1	25.9	30.3	28.3	31.3	38.3	34.9
Chi-square	11.844***	20.705***	30.029***	1.081	6.151	6.342*	9.456**	25.204***	32.316***

(continued)

Table 7.4 (continued)

	Rural			Urban			Total		
	Male (N = 114)	Female (N = 152)	Total (N = 266)	Male (N = 80)	Female (N = 102)	Total (N = 182)	Male (N = 194)	Female (N = 254)	Total (N = 448)
*Disability****									
No disability	25.0	31.3	28.1	22.0	19.7	20.8	24.1	27.5	25.8
At least one	33.5	40.0	36.9	23.8	26.4	25.1	29.9	35.2	32.6
Multiple disability	42.7	53.6	48.4	35.8	42.0	39.7	40.9	49.6	45.8
Total	34.0	42.3	38.3	26.3	30.2	28.4	31.5	38.1	35.0
Chi-square	9.952**	15.664***	25.918***	3.386	9.786**	13.379***	13.596***	24.694***	39.357***
*Word recall***									
Less than or equal to two	29.3	57.5	47.4	26.7	36.0	32.5	28.1	52.0	43.2
Medium	35.0	40.9	38.1	25.4	31.0	28.7	32.1	37.5	35.1
High	33.0	32.3	32.7	26.9	25.6	26.4	30.6	28.6	29.8
Total	34.0	42.3	38.3	25.9	30.6	28.5	31.4	38.1	34.9
Chi-square	0.59	9.772**	6.148	0.056	0.869	0.565	0.448	12.117***	7.861**
*Frequency of participating in religious activities****									
Never	34.8	44.3	39.8	24.9	30.0	27.7	31.8	39.6	36.0
Occasionally	34.2	33.3	34.0	36.4	41.2	38.5	35.0	35.5	35.2
Regularly	19	0.0	10.3	25.0	12.5	20.0	21.2	3.8	13.6
Total	34.0	42.3	38.3	26.0	30.2	28.3	31.5	38.2	35.0
Chi-square	2.207	14.409***	14.227***	1.354	2.165	2.728	1.976	13.668***	12.465***
*ADL****									
No assistance	33.0	40.3	36.8	25.4	30.0	27.9	30.7	36.8	33.9
Partial assistance	40.0	75.0	55.6	66.7	33.3	50.0	42.9	57.1	50.0

(continued)

Table 7.4 (continued)

	Rural			Urban			Total		
	Male (N = 114)	Female (N = 152)	Total (N = 266)	Male (N = 80)	Female (N = 102)	Total (N = 182)	Male (N = 194)	Female (N = 254)	Total (N = 448)
Full assistance	56.2	100.0	77.4	40.0	40.0	40.0	52.4	85.0	68.3
Total	33.9	42.4	38.3	26.3	30.2	28.4	31.5	38.3	35.0
Chi-square	3.799	22.945***	22.098***	3.104	0.249	2.106	4.859	20.212***	22.102***

Note the table has been prepared by authors after analysis of data from the UNFPA survey (2011)
*** $p < 0.001$, ** $p < 0.01$, * $p < 0.05$

shows that 68.2% of elderly women in the oldest old age group living in rural areas, and 60.0% of oldest old men living in urban areas experienced elder abuse. Hence, the age of the elderly showed a significant association with experience of abuse, as did gender and place of residence.

Marital Status

Marital status of elderly people as yet another important variable that determines elder abuse emerged from a review of literature in developing countries. It was observed that 78.6% of those elderly people who had never married or were separated or divorced reported experience of abuse, with 78.6% of rural elderly women with this marital status reporting experience of abuse. Widowed elderly people also reported a high prevalence of abuse (39.5%). Gender disaggregation shows that widowed women from rural areas were the most vulnerable as over 45% of them reported experience abuse. In this study, a significant association was observed between marital status and elderly people's experience of abuse.

Caste

Caste is an important factor that influences the lives of people in India. It was observed that in this study, caste showed a strong association with experience of abuse. Some 39.3% from the forward caste reported experience of abuse, with elderly women from this caste reporting the highest (42.5%) incidence of abuse. Elderly people from other backward classes reported the least incidence (26.0%) of abuse among all caste groups. Some 50.7% of elderly women from scheduled castes and scheduled tribes living in rural areas reported experience of abuse.

Level of Education

Various studies have shown a negative association of abuse with level of education of elderly people, and the same was observed in this study. It was observed that seniors who were illiterate reported a higher incidence of abuse which reduced with increase in educational level. Although this association was significant at aggregate level, it wasn't statistically significant for elderly women in either rural or urban areas. Educational level of over 8 years among elderly halved (21.2%) the incidence of abuse as compared to illiterate elderly people (40.0%). Association between work participation and experience of abuse was studied and it was observed to be non-significant at aggregate level as well as by place of residence. Work participation did not show any significance by disaggregating data of elderly by gender.

Number of Children

Children provide a support system for senior people at one extreme and are the prime perpetrators of abuse at the other; hence, association between number of children alive and elder abuse was explored. It was observed that the incidence of abuse increased with the increase in number of children living, but this association was not significant at aggregate level. Some 32.7% of seniors who had up to two children alive reported experience of abuse, while two-fifths of those who had five or more children reported experience of abuse. However in urban areas, this association was observed to be statistically significant where having more children led to a greater proportion of seniors reporting abuse, especially elderly women.

Asset Ownership

Ownership of assets was also explored for its association with experience of elder abuse. It was observed that a higher percentage of elderly people who did not own land (35.9%) reported experience of abuse compared to those who did own land (33.5%). Though this association was not observed to be significant at aggregate level, it did show significance for elderly people living in rural areas. It was observed that 47.2% of elderly people living in rural areas without any ownership of land reported experience of abuse, while only a little over one-third of elderly people who owned land did so. Land ownership was observed to be significantly associated with experience of abuse by older women in rural areas.

Wealth Quintile

Wealth has been an important factor that has been shown to have a bearing on experience of abuse in Indian society. It was observed that the highest proportion of elderly people (51.4%) belonging to the poorest wealth quintile reported experience of abuse, while elderly people from the richest wealth quintile reported the lowest incidence (21.4%) of abuse. This association was observed to be statistically significant at aggregate level as well as by place of residence and across gender. This variation was especially evident in the rural areas, where 52.1% of total elderly people from the poorest quintile reported experience of abuse as compared to 13.3% of elderly people reporting experience of abuse.

Health and Disability

Health and disability levels of elderly people have an association with experience of abuse, in terms both of poor health status and disability leading to increased vulnerability to abuse and vice versa. In this study we explored this association, and observed that poor general health as significantly associated with experience of abuse among seniors. Some 38.3% of elderly people who reported poor health status reported experience of abuse while 23.8% of elderly people reporting good health reported abuse. This association was observed to be statistically significant at aggregate level, as well as by place of residence and gender. Similarly disability increases the dependence of elderly people, which increases their vulnerability to abuse. This phenomenon was clearly evident from the findings of this study: 25.8% of all elderly people without any disability experienced abuse while 45.8% with multiple disabilities reported did so. This phenomenon was observed to be statistically significant irrespective of gender and place of residence. Ability to perform ADL reflects elderly people's level of dependence, which may have an association with experience of abuse. Some 33.9% of elderly people who did not require any assistance to undertake their ADL reported experience of abuse compared to 68.3% of elderly people who were fully dependent on others for undertaking ADL reporting experience of abuse. This association was statistically significant at aggregate level as well as among rural elderly people. It is important to state that all elderly women (100.0%) from rural areas who required full assistance to undertake ADL and 75.0% of those who required partial assistance reported experience of abuse, and these associations were observed to be statistically significant.

Cognitive Ability

Cognitive ability declines with age, which has been reported to be one of the factors that increases the vulnerability of elderly people to abuse. Word recall is an indicator which is usually used to assess cognitive ability of elderly people. In this study, it was observed that 43.2% of elderly whose word recall was less than two words experienced abuse as compared to 29.8% of elderly people who recalled all words. This association was observed to be significant at aggregate level as well as for elderly women in total and in rural areas.

Community Participation and Social Engagement

Community participation and social engagement have been observed to have an influence on elderly people's vulnerability to abuse. Frequency of participation in religious activities was studied to understand community participation of seniors in this study. It was observed that elderly people who regularly engaged in religious activities at community level reported the lowest incidence of abuse (13.6%) as compared to those who participated occasionally (35.2%) or never participated (36.0%) in such activities. This association was statistically significant at aggregate level and for seniors living in rural areas. On disaggregating this data by gender it was observed that this association was significant for elderly women at aggregate level as well as for elderly women living in rural areas.

Case Study 1

Sheela (name changed) is a 75-year-old female widow from Kondhawale village where she lived until her marriage, when she shifted to Khechar village to her in-laws' place. Her husband owned a small piece of land where they both used to work, and her husband also had a daily-wage job as a carpenter, and they managed. After her husband died, she had no support from her family members or from the community. As her daughter at the time was very small, she found it difficult as she had to take care of her daughter and make a day-to-day living. Her in-laws took away her husband's property, harassed and insulted her when she asked for her share. She felt helpless as she did not get support from society, due to village traditions of sons having rights to the property, which she did not have. She worked as a labourer on a farm and there was no financial support on that end. Helpless, she decided to return to her mother's place in Kondhawale, where she took a daily wage job as an agricultural labourer. Within her limited income, she took care of her daughter but was unable to educate her. Her brother helped her in her bad days, financially supporting her, and with his help she conducted her daughter's marriage. Her daughter is happily married and her son-in-law takes good care of her, financially supports her, and visits her; and his brother also provides her with whatever money he gets from his religious prayers. She feels bad because her relatives never provided financial or emotional support to her. With age, she has lost her ability to work, cannot contribute and is often unable to cook. Due to poor health, her daughter moved to the village and her son-in-law also took a job in the village so that her daughter could take care of her. Her daughter is very supportive, taking care of the household work and hers as well, and the grandchildren respect her. She does feel lonely at home when all the family members are at work as she has no-one to speak to, as she sits idle all day at home due to bodily weakness. She feels good that her daughter is taking care of her, but is concerned about her own daughter's old age. She believes that the government should provide financial support to the elderly in the village and that the elderly who do not have male members at home should be provided with employment.

Case Study 2

Pournimabai (name changed) is 72-year-old widow who lives in Pune. However, she belongs to Jalgaon district. When she was younger, she had to drop out of school to take care of her brother (who had met with an accident) while helping at a farm and doing housework. When she got married, her husband never used to work or take responsibility of the home, and he used to abuse Pournima verbally. But she never complained to her parents regarding the same. She worked on daily wages and had no son (they had one daughter). Her husband was an alcoholic, due to which they lost the land, and he used to abuse her both physically care of her daughter, but When her daughter shifted to Pune after marriage, she could not meet her due to lack of economic support. After her husband's death, she stayed with her daughter in Pune, and both her daughter and son-in-law took care of her. To financially assist them, she worked as a domestic help but she is less able to work due to age. She feels bad because she is not getting work because of old age, but cannot be dependent on her daughter's family, as they have limited income and are unable to meet her increasing medical expenses. She feels that if she had a son, the situation might have been different, but she prefers to stay with her daughter, or she has no wish to live, because her daughter's good deed.

These case studies illustrates how patriarchy pushes women in India into a state of deprivation, where they lose autonomy, power, status in the family and access to their own resources due to widowhood and the implicit abuse experienced. The case also depicts the hardships experienced by women during their life courses, and its implications for their health and quality of life. The case studies also reveal older women's expectations of government intervention to ensure social security under such circumstances.

Case Study 3

Ramdeen (name changed) is 70 years old, illiterate, and from a young age used to work as an agricultural labourer. After his wife passed away, he stayed with his nephew and nephew's wife. Due to old age and ill health, he became unable to work and hence had no income so was soon completely dependent on them. Since then his relatives (nephew and his wife) started ill-treating him, quarrelling with him and asking him to leave their house. As he has no savings or property, they continued to torture him physically and mentally until he decided to take shelter in a temple. He says, 'When I was physically active and was able to work there was no conflict in the family. Once I became physically weak and economically poor my relatives did not want me anymore. They are forgetting the fact that 1 day they are also going to become old like me'. He receives Rs. 200 as old-age pension, on which he is fully dependent and which he gets after a few months, and which is not sufficient to meet his basic needs. Age has led to increasing health problems and increased medical costs. People from the village provide him with food much of the time, and he often does not take a bath as the distance between the village and the source of water is very far. He knows that hygiene also impacts his health, but due to no support, he feels helpless. In addition, he mentioned that the lack of facilities in the village (such as proper transport) does not allow him to visit hospitals. He feels that if he had had children they might not have behaved like his nephew and things might have been different. He attributes the abuse he receives to having no family or property. He says that he would be very happy if there were to be an old-age home in the village, as he would be cared for and food would be provided alongside other basic necessities. He felt sad because he was not aware of any welfare schemes for the elderly, but expressed his happiness to the interviewer as for the first time in his life someone was talking with him for a long time and showing concern towards him. He mentioned that the government should take care of helpless and vulnerable seniors, and fulfil their basic and medical needs under various schemes, while NGOs should create awareness about various schemes and policies that benefit the elderly people in every village. He expressed his anger at the Gram Sevak and Panchayat members for not allowing elderly persons to participate in the Gram Sabha, ignoring their views and opinions.

This case illustrates the plight of a widower who has no children of his own and is dependent on his relatives to take care of him and meet his basic needs. He experiences abuse from them, and perceives that lack of family members as well as any assets has made him dependent on his nephew and his family. He perceives that his dependence is the main cause of elder abuse, experienced by him. Though he receives some pension from government social-security schemes, it is infrequent and insufficient to meet his basic needs. Hence, he has no option but to continue to live with his nephew and continue to suffer from abuse and mistreatment.

Discussion and Recommendations

The findings clearly suggest that a high proportion of elderly people in Maharashtra experience mistreatment, neglect and violence, mainly from their own family members. Elderly women are more vulnerable than elderly men, and elderly people who are fully dependent on others for ADL or have multiple disabilities were observed to be most vulnerable. Very clearly, age of the elderly people was positively correlated with experience of abuse; hence the oldest old are more vulnerable to experience of abuse, probably due to their increased dependence on family members for undertaking ADL, financial dependence, and medical expenditure. Given these vulnerabilities, seniors look to the government to intervene to ensure social security as well their basic needs. As the life expectancy of those aged 60 is increasing gradually in India, this changing socio-cultural context and family relations and diminishing filial piety are the major drivers of the increase in the incidence of elder abuse in the state. This changing context and increasing elder abuse have serious implications for their quality of life which need to be addressed by adequate policy and programmatic measures.

Building Stronger Inter-generational Solidarity

As family is the unit for survival and well-being of old people, it is important that we draw from the strength of family ties and social networks, and provide services to support and enhance the care-giving role of family members. Efforts are necessary to promote value education among school-going children and training family care-givers in care of the elderly. There is a need to design welfare programmes in such a way that they reach the old people among the poor strata of our society. Media (print, audio, and visual) should highlight the mutual benefits of inter-generational bonding.

Promoting a Positive Social Attitude in the Community to Combat Ageism

People's attitude towards the old is one of the explanatory variables of elder abuse. The prevailing stereotypes about the role and status of elderly people in society have a strong bearing on how they are treated in the family and community. Therefore, it is important to develop positive attitudes about the elderly and their care among members of the family, especially the young ones. An important strategy to prevent elder abuse in society is to reduce ole people's social isolation by socializing them with the outside community programmes.

Social Security Measures to Ensure Financial Independence of Seniors

Old widows, in view of their high dependency and large proportion among the total old population, are likely to be the largest group among the abused. Therefore, all widows (especially among the poorer sections of our society) have to be provided with widow pension benefits.

Strengthening Health-Care Systems

It is necessary to ensure that all health and social care workers are adequately trained to respond to the needs of both the abused and the abusers. The National Programme for Health Care of Elderly (NPHCE) provides preventive, curative, and rehabilitative services to elderly persons at various levels of the health-care delivery systems. This needs to be effectively implemented by prioritizing districts with higher proportions of elderly people in the population. Strengthening NPHCE will go a long way in reducing the financial burden for treatment of acute and chronic morbidities and hospitalization among seniors. The Rashtriya Swasthya Bima Yojana scheme needs to be expanded to ensure coverage of all BPL elderly households. There is also a need to train health-care providers on the concept, scope and dynamics of elder sexual abuse and offer suggestions for appropriate responses that are applicable across a wide range of health and social services.

Capacity-Building and Respite for Care-Givers

If the dependency burden of caring for the young and elderly is shared equally between men and women, the problems of old people can be minimized to a great extent. However, in reality, it is usually the women who are the main care-givers for them and this imposes great constraints on the women. Therefore, efforts should be made to see that the dependency burden is equally shared between men and women. Government needs design programmes to provide respite to carers.

Effective Implementation of the Maintenance and Welfare of Parents and Senior Citizens Act (MWPSCA) (2007)

Though the MWPSCA was enacted in 2007, there are no effective mechanisms which encourage identification, reporting, and official intervention in cases of elder abuse. There is a need to improve and strengthen laws to offer greater protection to the elderly from abuse and neglect.

Expanding the Research to Explore the Phenomenon of Elder Abuse

Because elder abuse is a growing problem in our society, research on this must be to enable evidence-based decision making by policy makers and programme implementers.

References

American Psychological Association. 2012. Elder abuse and neglect: In Search of Solutions. Retrieved from, http://www.apa.org/pi/aging/resources/guides/elder-abuse.pdf. Accessed August 29, 2016.

Bagga, A., and A. Sakurkar. 2011. Abuse and violence in the lives of older women: Its impact on their mental health. *Help Age India- Research and Development Journal* 17 (3): 7–17.

Bhat, A.K., and R. Dhruvarajan. 2001. Ageing in India: Drifting intergenerational relations, challenges and options. *Ageing & Society* 21: 621–640.

Chokkanathan, S., and A.E. Lee. 2005. Elder mistreatment in urban India: A community based study. *Journal of Elder Abuse and Neglect* 17: 45–61.

Devi Prasad, B. 2000. Elder abuse and neglect: A review of research and programmes. In *Gerontological social work in India*, ed. Murli Desai, and Siva Raju, 219–239. Delhi: B.R. Publishing Corporation.

Devi Prasad, B., and B. Vijaya Lakshmi. 2008. *Psychosocial consequences of elder abuse and neglect: A study of the implications of elder abuse to mental well-being among rural elderly of north costal Andhra Pradesh (Project report).* Vishakhapatnam: Department of Social Work, AU.

Gore, M.S. 1968. *Urbanization and family change.* Bombay: Popular Prakashan.

Gupta, Nidhi. 2015. Physical and psychological health of older women in urban India. *Indian Journal of Gerontology* 29 (1): 23–45.

Gupta, Nidhi. 2016. Development, elder abuse and quality of life: Older women in urban India. *International Journal on Ageing in Developing Countries* 1(2): 158–173. https://www.inia.org.mt/wp-content/uploads/2017/01/1.2-8-India-158-to-173-Final.pdf Accessed February 12, 2017.

Hoban, Sheila, and Kathleen Kearney. 2000. Emergency: Elder abuse and neglect. *The American Journal of Nursing* 100 (11): 49–50.

Kumar, V. 1997. Ageing in India—An overview. *Indian Journal of Medical Research* 106: 257–264.

Kumar, Vijay S. 1991. *Family life and socio-economic problems of the elderly.* Delhi: Ashish Publishing House.

Latha, K.S. 2008. Elder abuse and neglect: A review. *Indian Journal of Gerontology* 22 (3/4): 467–479.

Maintenance and Welfare of Parents and Senior Citizens Act. 2007. Accessed online at http://socialjustice.nic.in/writereaddata/UploadFile/Annexure-X635996104030434742.pdf on 4 April 2017.

National Center on Elder Abuse. 2003. *The National Elder Abuse Incidence Study.* www.aoa.gov/abuse/report.

Patel, Mamta. 2010. Crimes against the elderly. *Indian Journal of Gerontology.* 24 (3): 395–403.

Prakash, Indira Jai. 2001. Elder abuse: Global response and Indian initiatives. *The Indian Journal of Social Work* 62 (3): 447–463.

Rajan, S.I. 2004. *Chronic poverty among Indian elderly.* (Working Paper 17). New Delhi: CPRCIIPA.

Rao, K.V., B. Devi Prasad, and Ch. Avataramu. 2000. Attitudes of the student youth and middle-aged persons towards the elderly. *The Indian Journal of Social Work* 60 (1): 42–54.

Registrar General of India. 2011. *Census of India, 2011.* New Delhi: Government of India.

Saxena, D.P. 2006. *Sociology of aging: Global scenario and Indian perspective.* New Delhi: Concept Publishers.

Sebastian, D., and T.V. Sekher. 2011. Extent and nature of elder abuse in Indian families: A study in Kerala. *Help Age India-Research and Development Journal* 17 (3): 20–28.

Shah, A.M. 1999. Changes in the family and the elderly. *Economic and Political Weekly* 34 (20): 1179–1182.

Shah, Geeta, Rosamma Veedon, and Sabiha Vasi. 1995. Elder abuse in India. *Journal of Elder Abuse and Neglect (Special Issue)* 6 (3/4): 101–117.

Siva Raju, S. 2011. Situational analysis of elderly In India. In *Voice of the elderly in India*, ed. S. Siva Raju, 1–29. New Delhi: B.R. Publishing.

Siva Raju, S. 2013. Situational analysis of older women. In *Older women in India: The context, issues and concerns*, ed. P.K.B. Nayar, 278–301. New Delhi: The Women Press.

Siva Raju, S. 2000. Ageing in India: An overview. In *Gerontological social work in India*, ed. Murli Desai, and Siva Raju. Delhi: B. R. Publishing Co.

Siva Raju, S. 2002a. *Health status of the urban elderly: A medico-social study*. Delhi: B. R. Publishing Co.

Siva Raju, S. 2002b. Meeting the needs of the poor and excluded in India, situation and voices, the older poor and excluded in South Africa and India, UNFPA. *Population and Development Strategies* 2: 93–110.

Srinivas, S., and B. Vijayalakshmi. 2001. Abuse and neglect of elderly in families. *Indian Journal of Social Work* 62 (3): 465–478.

UNFPA. 2012. *Report on status of elderly in selected states of India, 2011*. New Delhi: UNFPA. http://india.unfpa.org/publications/report-status-elderly-select-states-india-2011. Accessed August 16, 2016.

United Nations. 2002. *Abuse of older persons: Recognizing and responding to abuse of older persons in a global context-* report of the Secretary General, Economic and Social Council: E/CN.5/PC/2.

Usha, B. 1997. The abused elderly. *Indian Journal of Medical Research* 106: 389–395.

Visaria, P. 2001. Demographics of ageing in India. *Economic and Political Weekly* 36: 1967–1975.

WHO and International Network for the Prevention of Elder Abuse (INPEA). 2001. *Elder abuse in the health care services in Kenya*. Research Report. Accessed online at INPEA http://www.who.int/ageing/projects/elder_abuse/alc_ea_arg.pdf.

Chapter 8
Prevalence of Mistreatment of Older Persons and Their Associated Psychological Distress in a Selected Population of North India

Muthuvenkatachalam Srinivasan and Sandhya Gupta

Background

The global population of older persons aged 60 and above is expected to double from 542 million in 1995 to about 1.2 billion in 2025 (Elder Abuse 2015). In India, the size of the elderly population (aged 60 years and above) is fast growing, although it constituted only 7.4% of the total population at the turn of the new millennium. It is projected that the elderly population will form 26.7% of the total population in the year 2050. There is a great need to focus on holistic policies and programmes to deal with ageing-related issues of ageing society (Rajan 2015). Old age refers to ages that approach or exceed the average lifespan of human beings (Kurz 2015). Older persons can be described according to a range of characteristics including chronological age, change in social role and changes in functional abilities (National Policy on Older Persons 1999). The older persons are categorized as young old (60–64), middle old (65–69), older old (70–74) and oldest old (75 and above) (Alter 2004; Disabled Persons in India 2003). The distinctive signs associated with old age comprise both physical and mental characteristics (Salokangas and Joukamaa 1999; Birren and Fisher 1995). Issues related to old age are multi-dimensional, being physical, mental, social and economic. Failing physical and mental health, economic insecurity, isolation, neglect, abuse, fear, boredom, lowered self-esteem, loss of control and lack of preparedness for old age are some of the common issues related to older persons which need attention (Problems of Elderly 2015). Older persons are more likely to experience stressful events such as bereavement, a decline in socio-economic status with retirement, or a disability. All

M. Srinivasan
Faculty, Institute of Liver and Biliary Sciences, New Delhi, India
e-mail: Muthu.venky@gmail.com

S. Gupta (✉)
Faculty, College of Nursing, AIIMS, New Delhi, India
e-mail: drsandhyag407@gmail.com

Mala Kapur Shankardass and S. Irudaya Rajan (eds.), *Abuse and Neglect of the Elderly in India*, https://doi.org/10.1007/978-981-10-6116-5_8

these factors can result in isolation, loss of independence, loneliness and psychological distress (Mental Health and Older Adults 2016). Social issues affect an older person's risk and experience of illness. Social factors influencing the health of older persons include availability of family care-giver, socio-economic status, living arrangements, financial status, work history, education, history of trauma, and older persons' own care-giving responsibilities towards their family (Older People's Health Issues 2016; Oliveira et al. 2012). There is an increasing awareness throughout the world that some older persons are at high risk of mistreatment (Loughlin and Duggan 1998). Mistreatment refers to abuse, neglect, exploitation and abandonment of elderly people by family members, spouse, relatives or any other formal or informal care-givers (Missing Voices: Views of Older Persons on Elder Abuse 2016).

Abuse against the older person is recognized as an important challenge to elderly health. The World Health Organization (WHO) defines elder abuse as 'a single or repeated acts, or lack of appropriate action, occurring within a relationship where there is an expectation of trust, which causes harm or distress to an older person' (Elder Abuse 2016). Abuse includes the deliberate infliction of physical or mental injury, sexual abuse, withholding of necessary food, clothing and medical care to meet the physical and mental needs of an elderly person by the care-giver who is responsible for the care, custody, or responsibility of an elderly person (The Council on Scientific Affairs 1987). Verbal and emotional abuse can include yelling, swearing and making insulting or disrespectful comments. Psychological abuse involves any type of coercive or threatening behaviour that sets up a power differential between the older adult and his or her family member or care-giver. Sexual abuse can range from sexual exhibition, sexual assault or molestation to rape. Sexual abuse includes inappropriate touching, forcing the person to watch pornography, photographing the person in suggestive poses, forcing sexual contact with a third party, or any unwanted sexualized behaviour. Sexual abuse is perhaps the most egregious but least reported type of elder abuse (Elder Abuse and Neglect 2016).

Neglect is the failure of a care-giver to provide the necessities of life to an older person, such as adequate food, shelter, clothing, medical care, dental care or personal hygiene. Neglect is the intentional or unintentional failure of a care-giver to meet the physical and emotional needs of an elderly person, and care-giver neglect is either passive or active (Kruger and Moon 1999; Kleinschmidt 1997).

Exploitation or financial abuse can range from misuse of an older person's funds to misappropriation. Financial or material abuse includes the illegal or improper use of the older person's property or finances. It includes fraud, taking money under false promises, forgery, forced property transfers, purchasing expensive items with the older person's money without that person's knowledge or permission, or denying the older person access to his or her own funds or home (Kurre 2004).

Abandonment refers to the act of abandoning an older person by a carer or guardian without justifiable reasons, which endangers or ruins the physical and psychological health. Staying alone on streets, in parks, malls for a prolonged period and being dirty frequently are considered to be indicators of abandonment of the elderly (Protecting Elders Against Neglect and Abandonment 2015).

Mistreatment of the elderly is increasingly being seen as an important problem and one that is likely to grow as many countries experience rapidly ageing populations (Elder Abuse 2015). Mistreatment can exacerbate an older person's chronic and disabling conditions of older person and make them more dependent, vulnerable, and marginalized. By and large, this serious social issue was hidden from public view until recently and considered mostly a private matter. Even today, elder abuse continues to be a taboo, mostly underestimated and ignored by societies across the world (Gupta and Chaudhuri 2008).

Elder abuse is a major issue in both developing and developed countries yet it is an underestimated and under-reported globally. Prevalence rates or estimates exist only in selected developed countries, ranging from 1 to 10%. Even though the exact prevalence and magnitude of elder mistreatment are unknown, its social and moral significance is noticeable (The Council on Scientific Affairs 1987). Most studies report that older persons experience multiple barriers to reporting mistreatment (Ahmad and Lachs 2002). Prevalence of elderly mistreatment in urban India was 14% and chronic verbal abuse was the most common type of abuse followed by financial abuse (Srinivasan and Lee 2005). Studies both overseas and in India suggest that between 4 and 10% of patients referred to aged care services are victims of abuse and neglect. Mistreatment experienced by an older person may be one of the biggest source of stress and distress, especially for those with few support resources. In addition, those who experience mistreatment may face greater chances of psychological distress, which may be diminished by their psycho-social resources and aggravated by negative features of their close relationships; the relationship of mistreatment with depression and poor health emphasized that older persons with depression symptoms and poor health should be screened for mistreatment (Srinivasan 2015; Luo and Waite 2011).

Elder mistreatment is an important public health problem. It is estimated that one in ten older persons experience abuse each month. This is likely to be under-reported, as only one in 24 cases of elder abuse is reported, in part because older persons are often afraid to report cases of abuse to family, friends, or to the authorities (Elder Abuse 2015).

Certain societal attitudes in our community may contribute to the lack of value and respect for older adults, such as the belief that what happens at home is a private family matter and it is private inflicted by their kith and kin for fear of being humiliated and further harm (Srinivasan 2015).

An explorative study conducted among older persons from rural Tamil Nadu (2015) to find out the relationships between psycho-social resources and deficits, elder mistreatment, and psychological well-being found that lower levels of positive support, higher levels of criticism from within close relationships, and feelings of social isolation are positively associated with self-reported mistreatment experience. Older adults with fewer psycho-social resources or more psycho-social deficits seem to be more vulnerable to mistreatment, and mistreatment seems particularly detrimental to psychological well-being for these people.

A study of 200 older patients attending the rural and urban geriatric clinics at a medical college hospital in Bangalore, India (2015) reported that 16% of the older persons experienced abuse. Verbal abuse, neglect, financial abuse and physical abuse were experienced by 12.5, 11, 8.5 and 1.5% of older persons respectively. Elder abuse had significant association with financial dependence, lack of social support, and depression of elderly patients (Nisha et al. 2015).

A cross-sectional survey was conducted on elderly abuse from representative populations aged 60 and above from seven Indian states across India (2014). Findings of the survey from a total of 9852 elderly from 8329 households from the states of Himachal Pradesh, Kerala, Maharashtra, Odisha, Punjab, Tamil Nadu, and West Bengal revealed that 11% of those aged over 60 have experienced at least one type of abuse (physical, 5.3%; verbal, 10.2%; economic, 5.4%; disrespect, 6%; neglect 5.2%). Sons of elderly people were reported to be responsible for the abuse of 41% of male victims and 43% of female victims (Skirbekk and James 2014).

An explorative study was conducted to determine the association between verbal mistreatment, depression and quality of life among elderly people in the USA. Some 38% of 142 older adults (40% male) reported verbal mistreatment, and the study also found that verbal mistreatment was a significant predictor of social functioning ($r = -0.28$, $p < 0.001$), mental health ($r = -0.25$, $p < 0.001$), and role limitations OR = 3.02, 95% (Fulmer et al. 2014).

A study conducted by HelpAge India among 5600 older persons across the country, including 20 census blocks, using cluster sampling to find out the scale and expanse of elder abuse in the country revealed that 31.13% of older persons reported abuse. Madhya Pradesh reported the highest prevalence with 77.12%, followed by Assam (60.55%) and Uttar Pradesh (52%). Older persons from Uttarakhand state reported a 13.31% prevalence of abuse while Rajasthan reported the least prevalence at 1.67%. The only two states which reported prevalence in single digits were Rajasthan (1.67%) and Himachal Pradesh (2.99%) (HelpAge India 2012).

A cross-sectional survey was conducted among older adults aged 60 or above in three rural communities of China to determine mistreatment and its correlates. Data was collected through a structured questionnaire from 2039 participants and the findings revealed that mistreatment was reported by 36.2% of participants. Prevalence rates of psychological mistreatment, care-giver neglect, physical mistreatment, and financial mistreatment were 27.3, 15.8, 4.9 and 2.0% respectively. Depression, being widowed/divorced/single/separated, having a physical disability, having a labour-intensive job, depending solely on self-made income, and living alone were identified as risk factors for elder mistreatment (Wu et al. 2012).

A descriptive survey examined the prevalence and severity of various forms of abuse and neglect from the victims' perspective (2009). Data was collected through personal interviews from a national representative sample of community urban dwellers age 65 and above, using cluster sampling techniques and sampling proportionately both Arab and Jewish elders. Findings indicate that 18.4% of the respondents were exposed to at least one type of abuse during the 12 months preceding the interview, the highest form being verbal abuse followed by financial exploitation. Women were more exposed to physical violence. Financial exploitation was mostly inflicted by adult children. Neglect in answering primary needs was found among 20% of the sample, regarding the 3 months preceding the interview (Lowenstein et al. 2009).

In the absence of data on mistreatment of older persons, elder abuse and its detection are challenging and highly sensitive issues that need a linguistically and culturally specified approach and vocabulary. There is a dearth of culturally sensitive tools to detect all four aspects of mistreatment of older persons without asking direct and sensitive questions. This study is aimed at identifying the prevalence of mistreatment among older persons and assessing their psychological distress. Identification of cases of abuses among elderly people by appropriate screening tests and prompt reporting of mistreatment to service providers may help to provide community services to the victims and prevent negligence and abuse.

The elderly were screened for mistreatment which includes abuse, neglect, mistreatment and abandonment and their psychological distress to examine whether experience of mistreatment coexisted with psychological distress, and whether the role of social support available to older persons had any beneficial effect in a selected community of Uttarakhand, taking into account their socio-demographic variables.

- **Prevalence of mistreatment** refers to the number of cases of mistreatment reported by older persons, which are expressed as per hundred old persons in that population block.
- **Mistreatment** refers to abuse, neglect, exploitation, and abandonment of elderly people by family members/spouses/relatives. The older person is identified as being mistreated if they reported at least one of the four subtypes of mistreatment.

  Abuse: includes intentional infliction of physical or mental injury, sexual abuse, or withholding of necessary food, clothing, and medical care to treat the physical and mental health needs of an older person by one having the care, custody or responsibility of an older person.
  Neglect: refers to failure to fulfil other needs which the victim is unable to provide for themselves.
  Exploitation: refers to misuse of a vulnerable older person's property, income, or other financial resources.
  Abandonment: involves a desertion of older persons or withdrawal of care without alternative arrangements by the family, spouse, or relatives.

- **Psychological distress** refers to an aversive state in which a person is unable to adapt completely to stressors and their resulting stress, and shows maladaptive behaviour. It includes psychological distress of older persons as measured by the scores of Kessler's psycho-social distress scale (K10), and Social support. Social support is measured by the scores of MOS Social Support Scale (MOS-SSS).

The Setting

Data for the present study was collected from Uttarakhand state in Northern India: seven villages in Haldwani Block, Nainital District of Uttarakhand state, with a population of 10,086,292 (2011), having 5,137,773 (50.94%) males and 4,948,519 (49.06%) females. The percentage of elderly people in the state is 7.5% (Uttarakhand Population Census Data 2011). Nainital district is located in Kumaon division of Uttarakhand, having a population of 955,128 (2011) and a population density of 225 people per km^2 (Nainital District 2011). Haldwani block has both urban and rural populations. The seven studied villages (out of a total of 218) were Khera, DewalaTalla, DewalaMalla, Dewala Talla Simlar, NawarKhera, KishanNagri, and Gaulapar.

Older Persons Under Study

All 381 older persons who were available in the selected villages during the data collection period of January–November 2015 were included in the study.

Characteristics of Older Persons

The mean number of family members of older persons was 7.46 (SD ± 3.56). Out of 381 older persons, 54.9% were female, 78.2% were married and 20% of them were widows or widowers. Most (97.1%) of the older persons belonged to the Hindu religion and most (61.9%) were living in a joint-family. Most (63%) of the older persons were living with a male child and 26.2% of them were living with more than one child in the same house. Most of the older persons (63.5%) were living permanently with the same child, nine older persons (2.3%) kept on shifting from one child's family to another as per their wishes to change their place of stay, and only one older person (0.3%) shifted their living arrangements on a monthly basis.

A little over half (53.5%) of the older persons had no formal education and 25.7% had attended primary school. A substantial proportion of older persons (62.7%) had a monthly family income of Rs. 5000/- or less, and 6% had Rs. 15,001/- and above. Most of the older persons (78.5%) were still earning money either in the form of a pension (60.9%), self-employment (7.7%), interest from savings (1.7%) or having income from agriculture (29.7%).

A little over half (55.2%) of the older persons has six to ten family members, whereas 13.6% had 11 or more family members. Nearly half (49.6%) of the older persons had three to four children and 8.4% of them had seven or more. More than one-third (36.0%) of the older persons had two sons and 32.3% had three sons or more. Some 76.9% of the older persons had at least one married son or more (Table 8.1).

One-third (33.9%) of the older persons had only one daughter and one 24.7% had three or more, whereas 16.0% had no daughter. Two-thirds (66.1%) of had at least one married female child or more. Some 77.2% of the older persons five or fewer grandchildren and 19.4% had between 6 and 10.

Table 8.1 Socio-demographic characteristics of older persons (n = 381)

	Mean	SD
Age of the participant	68.17	7.64
Number of family members	7.46	3.56
	Frequency	**Percentage**
Gender		
• Male	172	45.1
• Female	209	54.9
Marital status		
• Married	298	78.2
• Never married	06	01.6
• Widowed	76	19.9
• Separated/divorced	01	0.3
Religion		
• Hindu	370	97.1
• Muslim	11	2.9
Type of family		
• Joint family	236	61.9
• Nuclear family	133	35.0
• Extended family	12	03.1
Presently living with		
• Male child	240	63.0
• Female child	12	03.2
• Jointly living	100	26.2
• Other relatives	06	01.6
• Alone	23	06.0
Stay permanently with same child's family		
• Yes	242	63.5
• No	10	02.6
• Not applicable	129	33.9
How often shifting from one family to another		
• Monthly	1	0.3
• Whenever I wish/choice is mine	9	2.3
• Not applicable	371	97.4

(continued)

Table 8.1 (continued)

	Mean	SD
Educational status		
• No formal education	204	53.5
• Primary school	98	25.7
• Middle school	40	10.5
• Metric	22	05.8
• Inter/diploma	12	03.2
• Graduate or more	05	01.3
Monthly family income		
• Rs. 5000/- or less	239	62.7
• Rs. 5001–10,000/-	62	16.3
• Rs. 10,001–15,000/-	57	15.0
• Rs. 15,001/- and above	23	06.0

Four standardized tools were selected after extensive review of related literature. Permission to use the standardized tools was obtained from the copyrighted author/publisher of the respective tool. The details of the tools are given below. *The MOS Social Support Survey (MOS-SSS)* was used to assess the social-support availability of older persons on four subdomains, semi-structured interview ($r = 0.87$). *Kessler Psychological distress scale (K10)* was used to assess the psychological distress of older persons, semi-structured interview ($r = 0.93$). *Hwalek-Sengstock Elder Abuse Screening Test (H-S/EAST)* was used to assess the risk of abuse of older persons, semi-structured interview ($r = 0.84$). Elder Assessment Instrument (EAI) was used to assess the mistreatment of older persons including four subtypes, i.e. abuse, neglect, exploitation and abandonment, semi-structured interview ($r = 0.83$).

All the older persons were informed about the study in detail, an information sheet was given out, and only those who were willing to participate were enrolled into the study. Informed written consent was taken prior to data collection from the older persons, and their confidentiality and anonymity were assured.

Making Contact

The older persons were contacted after gaining permission from the respective Panchayat head. A rapport was built with the residents of the village as most of them attended health camp and health-related issues were discussed through health talk and exhibition. The family members of the older persons were requested to allow the interview in privacy as the data was being collected on a health-related issue, on which most of the families were cooperative and direct questions related to mistreatment were not asked in front of them. On the first day, after obtaining socio-demographic details, the social support was assessed and the same older persons were visited a second time and the rest of the data collection was completed to assess psychological distress, anxiety, depression, risk of abuse and mistreatment through semi-structured interview and observation. It took approximately 75–90 min to complete the data collection.

Findings and Discussion

Description of Older Persons

There were 381 older persons, the mean age being 68.17 years (SD = 7.64), with ages ranging between 60 and 98 years. More than half of the older persons were female (54.9%). More than three-quarters (78.2%) were married and 20% of them were widows/widowers. A little over half (53.5%) had no formal education. A substantial proportion (62.7%) had a monthly family income of Rs. 5000/- or less. The findings of these demographic characteristics are in conformity with other studies by Srinivasan (2015), Joseph et al. (2015), Dongre and Deshmukh (2012), and Shivakumar et al. (2015). Most of the older persons were Hindus (97.1%) and most (61.9%) lived in a joint-family. Satheesh et al. (2015) and Sampath Kumar et al. (2015) also reported similar findings on religion and type of family. The present study found that most (63%) of the older persons were living with a male child. Similar findings were reported by Kumar et al. (2015), but in contrast to the present findings Kumar et al. (2015) reported that only 30% of the older persons stayed with a son and 20% with a daughter.

Prevalence of Mistreatment Among Older Persons in a Selected Community of Uttarakhand

Prevalence of Mistreatment

Older persons are identified as having been mistreated if they reported at least one sub-type (i.e. abuse, neglect, exploitation or abandonment). As shown in Fig. 8.1, 15.7% of the older persons reported mistreatment and most (84.3%) did not report it. Out of 381 older persons, 60 reported at least one sub-type of mistreatment and the prevalence of mistreatment of older persons was 15.7% (95% confidence interval [CI] 12.05–19.35%). This finding is supported by HelpAge India (2012), Lowenstein et al. (2009), Nisha et al. (2015), and Srinivasan (2015) reported similar findings on the prevalence of mistreatment of older persons. Findings by Skirbekk and James (2014) and Saikia et al. (2015) reported a marginally lower prevalence of mistreatment of elderly people at 11 and 9.3% respectively. Compared with the present study, findings by Srinivasan (2015), Lowenstein et al. (2009), HelpAge

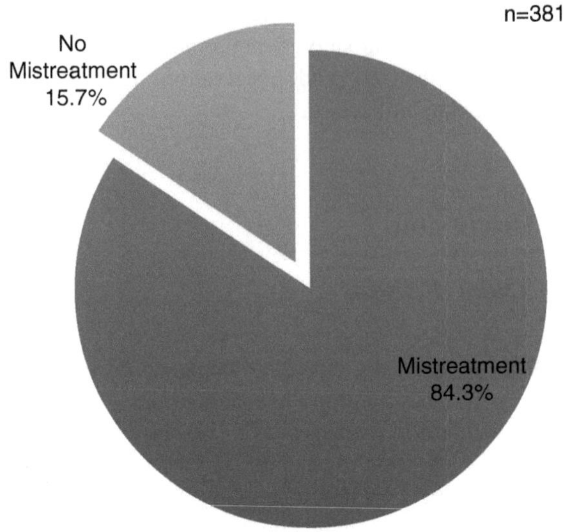

Fig. 8.1 Pie diagram showing mistreatment reported by older persons

Table 8.2 Prevalence of mistreatment among older persons according to sub-types (n = 381)

Sub-type of mistreatment	n	Prevalence (%)	Margin of error (at 95% CI) (%)	Prevalence (at 95% CI) (%)
Mistreatment	60	15.7	3.65	12.05–19.35%
• Abuse	19	05.0	2.19	2.81–7.19%
• Neglect	48	12.6	3.33	9.27–15.93%
• Exploitation	22	05.8	2.35	3.45–8.15%
• Abandonment	14	03.7	1.90	1.80–5.60%

Note Margin of error (precision) was calculated for prevalence of elder mistreatment and sub-types based on sample size (n = 381) at 95% confidence interval (CI)

India (2011), Martins et al. (2014), Dong et al. (2007), Carmona-Torres et al. (2015), and Abdel Rahman and El Gaafary (2011) reported a much higher prevalence of elder abuse ranging from 21 to 43%.

As shown in Table 8.2, 15.7% (95% CI 12.05–19.35%) of older persons reported mistreatment out of whom 5% (95% CI 2.81–7.19%) reported abuse, 12.6% (95% CI 9.27–15.93%) reported negligence, 5.8% (95% CI 3.45–8.15%) reported exploitation and 3.7% (95% CI 1.80–5.60%) reported abandonment. The overall prevalence of mistreatment is 15.7%. The most common type of mistreatment reported by older persons was neglect (12.6%), followed by exploitation (5.8%), abuse (5%), and abandonment (3.7%).

Prevalence of Abuse

In the present study, abuse, a sub-type of mistreatment, was reported by 19 older persons. The prevalence of elder abuse was 5.0% (95% CI 2.81–7.19%). Similar findings were reported by Abdel Rahman and El Gaafary (2011) and Acierno et al. (2010). Findings of the study by Amstadter et al. (2011), Fulmer et al. (2014), and Wu et al. (2012) on prevalence of abuse reported higher a prevalence, while Dong (2009) and Naughton et al. (2011) reported a lower prevalence (1.2 and 2.2% respectively) than the present study.

Prevalence of Neglect

Some 48 older persons reported neglect, and the prevalence of neglect was 12.6% (95% CI 9.27–15.93%). Similar findings were reported by Nisha et al. (2015) and Martins et al. (2014) with a prevalence of neglect ranging from 11 to 16.6%. Amstadter et al. (2011) conducted a study on mistreatment in South Carolina and reported a prevalence of 5.4%, which is lower than the 12.6% prevalence found in the present study.

Prevalence of Exploitation

Some 22 older persons reported exploitation and the prevalence of exploitation was 5.8% (95% CI 3.45–8.15%). Similar findings were reported by Acierno et al. (2010) and Amstadter et al. (2011) (5.2 and 6.6% respectively). A cross-sectional study conducted by Abdel Rahman and El Gaafary (2011) in Egypt reported slightly lower prevalence of exploitation (3.8%) than the present study.

Prevalence of Abandonment

Some 14 older persons reported abandonment and the prevalence of abandonment of older persons is 3.7% (95% CI 1.80–5.60%). Most of the studies on mistreatment did not report abandonment of elderly people in particular. Studies assessed and reported abandonment of older persons combined with neglect (Dongre and Deshmukh 2012; Joseph et al. 2015).

As given in Fig. 8.2, out of 15.7% older persons who reported mistreatment, 7.3% reported only one type of mistreatment, i.e. abuse or negligence or exploitation or abandonment, 6% reported at least two types of mistreatment, 1.8% reported any three types of mistreatment, and only 0.5% reported all four types of mistreatment.

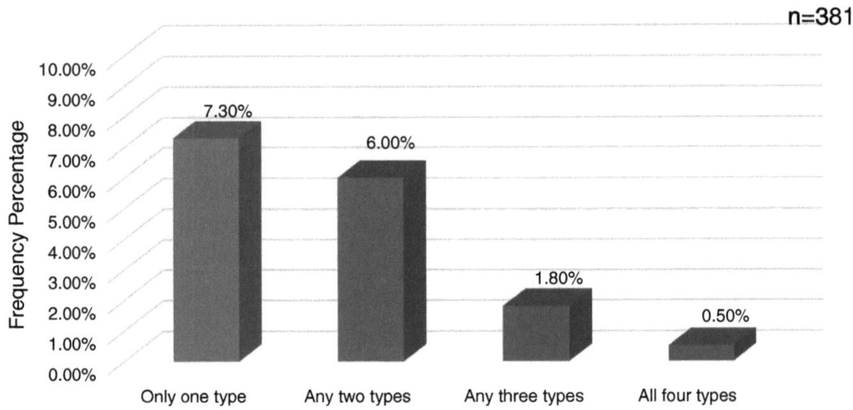

Fig. 8.2 Bar graph showing coexistence of sub-types of mistreatment among older persons. *Note* Four sub-types of mistreatment includes abuse, negligence, exploitation, and abandonment

Table 8.3 Coexistence of four sub-types of mistreatment as reported by older persons

Type of mistreatment of older persons	Frequency	Percentage
• Abuse only	08	2.1
• Negligence only	17	4.5
• Exploitation only	03	0.8
• Abandonment only	00	0.0
• Abuse + negligence	04	1.1
• Abuse + exploitation	00	0.0
• Abuse + abandonment	00	0.0
• Negligence + exploitation	09	2.4
• Negligence + abandonment	09	2.4
• Exploitation + abandonment	01	0.3
• Abuse + negligence + exploitation	05	1.3
• Abuse + exploitation + abandonment	00	0.0
• Abuse + negligence + abandonment	00	0.0
• Negligence + exploitation + abandonment	02	0.5
• All four type of mistreatment	02	0.5

As given in Table 8.3, only 0.5% of old people reported all four sub-types of mistreatment, 4.5% reported only negligence, 2.1% reported only abuse and 0.8% reported only exploitation. Thus, it can be interpreted that among those nine older people who reported mistreatment (n = 60), more than half (53.5%) reported multiple (more than one sub-type) types of mistreatment. Out of 381 older persons, 60 reported at least one type of mistreatment, i.e. abuse, neglect, exploitation and abandonment. A study conducted in rural areas of India by Srinivasan (2015) found

Table 8.4 Association of mistreatment reported by older persons and risk of abuse (n = 381)

Variables of older persons	No. of older persons		DF	χ^2 value	p value
	Reported mistreatment f (%)	Did not report mistreatment f (%)			
Older persons with					
• Risk of abuse	51 (52.0%)	47 (48.0%)	1	131.0	<0.001
• No risk of abuse	09 (3.2%)	274 (96.8%)			

that 83.4% of abused individuals showed simultaneous occurrence of multiple types of mistreatment which is in contrast to the findings of the present study.

Association of Mistreatment of Older Persons with Risk of Abuse

As shown in Table 8.4, more than half (52%) of the older persons who are at risk of abuse reported mistreatment, whereas only 3.2% of older persons who are not at risk of abuse reported mistreatment. There is a significant association (χ^2 = 131.0, $p < 0.001$) between mistreatment and risk of abuse.

Assessment risk of abuse revealed that a quarter (25.7%; 95% CI 21.4–30.0) of the older persons are at risk of abuse. Thus, it can be inferred that the older persons who are at risk of abuse are more likely to be mistreated than those who are not at risk of abuse. The findings of Strasser et al. (2013) are consistent with the present study; and in contrast, Saikia et al. (2015) reported 9.3% are at risk of abuse.

Psychological Distress and Social Support Among Older Persons

Presence of Psychological Distress in Older Persons

The psychological distress score is categorized into four categories as normal (below 20), 'mild' (20–24), moderate (25–29) and severe (30 and above). As shown in Fig. 8.3, a little over a quarter (27.3%) of older persons have mild psychological distress, 3.7% of them have moderate and only three older persons (0.8%) have severe psychological distress.

Thus it can be interpreted that nearly one-third of the older persons (31.8%) have mild, moderate or severe psychological distress and nearly one in 20 older persons (4.5%) have moderate to severe psychological distress. More than a quarter (27.3%) of older persons have mild distress, 3.7% of them have moderate distress, and three older persons (0.8%) have severe psychological distress. A retrospective study by

Shivakumar et al. (2015) in Bangalore revealed that 50% of the older persons who were seeking care for neuropsychiatric conditions had psychological distress. A study conducted by Atkins et al. (2013) among older persons in Australia found that psychological distress was experienced by 15% of the residential sample and 7% of the community sample, which was smaller than the 27.3% psychological distress reported in the present study.

Assessment of the Social Support Available to Older Persons

The availability of social support for older persons is assessed through MOS social support scale in four sub-domains. The score of each domain is calculated and converted to a scale of 0–100 as per the instruction of the tool.

As given in Table 8.5, the overall social support availability of older persons is 64.3 (SD = 21.6) out of a possible 100, and tangible support availability (66.4%) is better than the other three domains of social support, i.e. affectionate support

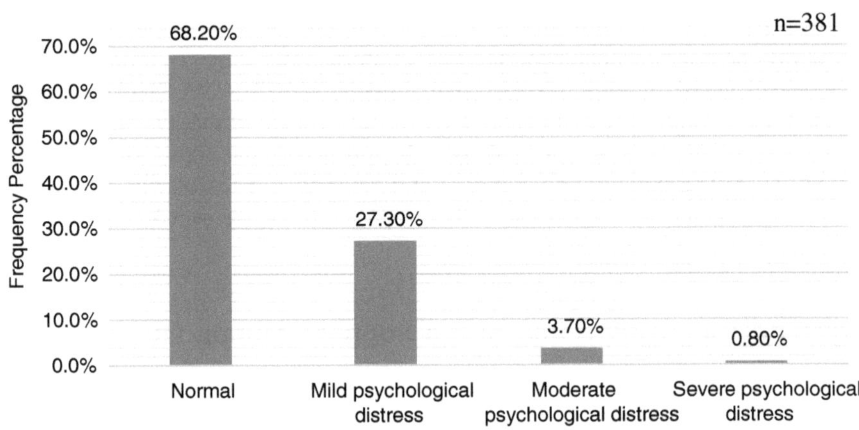

Fig. 8.3 Bar graph showing the older persons having psychological distress of mild, moderate and severe categories

Table 8.5 Domain-wise mean social support scores of older persons (n = 381)

Social support domains	Social support scores of older persons		
	Mean ± SD	Median	Range
Emotional/Informational Support (0–100)	62.3 ± 22.3	65.6	0–100
Tangible support (0–100)	66.4 ± 20.3	68.8	0.8–100
Affectionate support score (0–100)	64.3 ± 24.6	66.7	0–100
Positive social Interaction (0–100)	64.1 ± 23.5	66.7	0–100
Overall support score (0–100)	64.3 ± 21.6	66.7	0–100

(64.3%), positive social interaction (64.1%) and emotional or informational support (62.3%). In the present study, the tangible support available to the older persons is better than the other three domains of support, while the emotional/informational support score is lowest. The overall support score is 64.3 (SD = 21.6), which is nearly two-thirds of the maximum possible score (100) on the support scale. Similar findings were reported by Sharoni et al. (2015) and Gallegos-Carrillo et al. (2009).

The Relationship Between Mistreatment of Older Persons and Their Psychological Distress Score and Availability of Social Support

Association of Mistreatment of Older Persons with Their Psychological Distress

As given in Table 8.6, mistreatment, abuse, neglect, and exploitation of older persons have a statistically significant association ($p < 0.05$) with their psychological distress whereas abandonment did not have any association with their psychological distress score. Thus, it can be interpreted that the older persons who reported mistreatment are more likely to have psychological distress than those who did not report it.

In the present study, the older persons who reported mistreatment have significantly higher psychological distress scores than those who did not report

Table 8.6 Association of mistreatment of older persons and their psychological distress score (n = 381)

Mistreatment reported by older persons		*n*	Psychological distress score of older persons		*z* value	*p* value*
			Median (IQR)	Mean rank		
Mistreatment	No	321	17.0 (14–20)	182.8	3.37	0.001
	Yes	60	21.0 (15–23)	234.8		
Sub-types of mistreatment						
Abuse	No	362	17.0 (14–20)	187.0	3.12	0.002
	Yes	19	22.0 (16–24)	267.5		
Neglect	No	333	17.0 (14–20)	184.8	2.89	0.004
	Yes	48	20.5 (15–23)	233.8		
Exploitation	No	359	17.0 (14–20)	188.1	2.12	0.034
	Yes	22	21 (15–22.5)	239.2		
Abandonment	No	367	17 (14–21)	190.6	0.35	0.729
	Yes	14	17 (14.8–21)	201.0		

IQR Interquartile range
*Mann–Whitney '*U*' test was used

mistreatment. The findings of the study conducted by Luo and Waite (2011) are consistent with the findings of the present study. Both the studies reported that the mistreatment of older persons was positively associated with higher psychological distress.

Association of Mistreatment of Older Persons with Availability of Their Social Support

As shown in Table 8.7, mistreatment, neglect, and exploitation of older persons are significantly ($p < 0.05$) associated with their social support. Hence, it can be interpreted that older persons who reported mistreatment, neglect, or exploitation are likely to have less social support than those who did not report. Abuse and abandonment of older persons were not associated with their social support score.

In the present study population it was found that the older persons who reported mistreatment, neglect, or exploitation are likely to have poorer social support that those who did not report mistreatment, which were statistically significant. Wu et al. (2012), Fulmer et al. (2014), and Nisha et al. (2015) also reported similar findings.

Table 8.7 Association of mistreatment of older persons with their social support scores (n = 381)

Mistreatment reported by older persons		*n*	Social support score		*z* value	*p* value*
			Median (IQR)	Mean rank		
Mistreatment	No	321	67.4 (54.2–79.6)	197.2	2.54	0.011
	Yes	60	60.2 (38.7–75.6)	157.8		
Sub-types of mistreatment						
Abuse	No	362	66.9 (52.9–79.8)	192.9	1.45	0.148
	Yes	19	61.2 (46.1–72.7)	155.4		
Neglect	No	333	67.4 (53.9–79.4)	195.3	2.01	0.044
	Yes	48	61.0 (35.2–76.5)	161.1		
Exploitation	No	359	67.4 (53.9–79.7)	196.2	3.71	<0.001
	Yes	22	43.7 (30.1–65.5)	106.4		
Abandonment	No	367	66.9 (53.4–79.4)	192.8	1.65	0.1
	Yes	14	53.2 (29.2–74.6)	143.5		

IQR Interquartile range
*Mann–Whitney '*U*' test was used

Association of Mistreatment of Older Persons with Their Socio-demographic Variables

As shown in Table 8.8, mistreatment of older persons has significant association ($\chi^2 = 8.0$, $p = 0.018$) with age, and the older persons who aged 80 and above being more likely to be mistreated than those who are younger than them. Half (50%) of the older persons who were staying with relatives and 43.5% of older persons who

Table 8.8 Association of mistreatment of older persons with socio-demographic variables ($n = 381$)

Socio-demographic variables	Number of older persons		*df*	χ^2 value	p* value
	Did not report mistreatment *f* (%)	Reported mistreatment *f* (%)			
Gender					
• Male	144 (83.7%)	28 (16.3%)	1	0.067	0.796
• Female	177 (84.7%)	32 (15.3%)			
Age group					
• 60–69 years	210 (87.0%)	28 (13.0%)	2	12.44	0.002
• 70–79 years	85 (81.7%)	19 (19.5%)			
• ≥ 80 years	26 (66.7%)	13 (33.3%)			
Marital status					
• Married	254 (85.2%)	44 (14.8%)	1	0.996	0.311
• Never married/widowed/divorced	67 (80.7%)	16 (19.3%)			
Religion					
• Hindu	311 (84.1%)	59 (15.9%)	1	–	1.0[a]
• Muslim	10 (90.9%)	1 (9.1%)			
Type of family					
• Joint family	205 (86.9%)	31 (13.1%)	1	3.190	0.074
• Nuclear family	116 (80.0%)	29 (20.0%)			
Monthly family income					
• Rs 5000/- or less	197 (82.4%)	42 (17.6%)	3	5.35	0.148
• Rs 5001–10,000	50 (80.6%)	12 (19.4%)			
• Rs 10,001–15,000	53 (93.0%)	04 (7.0%)			
• Rs 15,001 and above	21 (91.3%)	02 (8.7%)			
Income status of elderly					
• Still earning	253 (84.6%)	46 (15.4%)	1	0.13	0.710
• Not earning	68 (82.9%)	14 (17.1%)			

(continued)

Table 8.8 (continued)

Socio-demographic variables	Number of older persons		df	χ^2 value	p* value
	Did not report mistreatment f (%)	Reported mistreatment f (%)			
Educational qualification					
• No formal education	169 (82.8%)	35 (17.2%)	3	2.67	0.445
• Primary school	81 (82.7%)	17 (17.3%)			
• Metric	55 (88.7%)	07 (11.3%)			
• Inter/diploma/ degree	16 (94.1%)	1 (5.9%)			
Number of children					
• 0–2	65 (81.1%)	17 (18.9%)	2	2.47	0.291
• 3–4	164 (86.8%)	25 (13.2%)			
• More than 4	92 (82.2%)	18 (17.8%)			
Presently living with					
• Son	205 (85.4%)	35 (14.6%)	4	17.70	0.001
• Daughter	9 (75.0%)	3 (25.0%)			
• Jointly living	91 (91.0%)	9 (9.0%)			
• Other relatives	3 (50.0%)	3 (50.0%)			
• Alone	13 (56.5%)	10 (43.5%)			

Note Chi-square test
[a]Fisher exact test

were staying alone reported mistreatment. Mistreatment of older persons has a significant association (χ^2 = 17.7, p = 0.001) with their living arrangements. Thus, it can be inferred that the older persons living with relatives or alone are more likely to be mistreated than those who live with a son, daughter or together (jointly living). Mistreatment of older persons has no association ($p > 0.05$) with their gender, marital status, religion, type of family, monthly family income, educational status and number of children.

In the present study it was found that older persons who are aged 80 and above are more likely to be mistreated than those who are younger. Wang et al. (2006) and Saikia et al. (2015) reported that age is negatively associated with elder abuse, which is consistent with the findings of the present study. The present study found that older persons living with relatives or alone were more likely to be mistreated than those who lived with their son, daughter, or together (jointly living). Similar findings were reported by Saikia et al. (2015), Martins et al. (2014), and Dong (2009).

Implications for Taking Immediate Steps

- *For older persons*: Empowerment through self-education. Optimizing opportunities for health, participation, and security in order to enhance quality of life of for people as they age. Proper retirement planning will help older persons retire with a positive frame of mind and enough coping skills to adjust to the demands of self, family, and society. Every individual's self-development can occur with healthy lifestyle, complementary therapies, and empowering activities.
- *For families*: Valuable education to family care-givers of older persons should be offered to develop positive attitudes to tackling mistreatment. Children in the family should be taught that respecting and taking care of elderly people is their responsibility. Family care-givers should be made aware about law enforcement regarding the Maintenance of Welfare of Parents and Senior Citizens Act, 2007. Counselling to care-givers of older persons with behavioural or personal problems will have a positive impact on preventing mistreatment of them.
- *For community and society*: Promoting positive social attitudes is the key to preventing mistreatment of older persons. Educating people in the community about elder abuse and promoting social contact should be done through campaigns. People in the community must be made aware that they should not be silent spectators when they witness abuse and they should intervene personally or report it to the relevant authorities. Mobilizing social contacts and societies involving peer groups for recreational activities will keep the older persons active and productive. Communities can develop their own programmes that are structured around meeting the needs of older persons.
- *For health-care providers*: Health-care providers at periphery level should be trained on assessment and identification of elder mistreatment. Health-care providers should be educated about laws on elder abuse so that they are aware and confident about interventions. Health-care providers should recognize the needs of older persons and their family care-givers, and plan personalized preventive strategies to tackle elder mistreatment.
- *For physicians*: Suspected cases of mistreatment when older persons visit their clinic for treatment should be investigated, and if mistreatment is found it should be reported.
- *For nurses*: Whenever they are performing physical examination of an elderly patient, the nurse as a primary-care provider should look for signs of physical abuse, sexual abuse, and neglect. Health education of family members regarding ageing and care of the elderly is necessary to encourage a positive attitude among family members towards mistreatment of older persons. Health education regarding the impact of elder mistreatment and the importance of avoiding elder mistreatment can be routinely given to family care-givers present in geriatric ward settings in hospital.

- *For social workers*: While performing their routine functions as a social worker they should look for signs of abuse and neglect of older persons and provide the education to community and family members regarding ageing and care of elderly necessary to improve family members' positive attitude towards tackling mistreatment. Counselling services can be provided to family members of older persons who are victimized and the information regarding legal provisions and other welfare services can be provided.

Implications for Government

- *At local level*: Local Panchayats can set up recreational facilities specially earmarked for older persons so that they will have the chance to interact and build social relationships among their peer groups. Self-help groups for older persons with members of retired and educated older persons can be formed with an aim to address problems of victims of elder mistreatment. The Panchayat can maintain statistics of incidents of elder mistreatment within its jurisdiction and report to government from time to time.
- *At state level*: Though there are law enforcements and legal provisions to protect older persons, there is a need to implement these measures vigorously to curb elder mistreatment.
- *Implications for police*: Police headquarters of each district should have dedicated cells for older persons with a toll-free helpline number. Police should widely publicize this helpline number, ensuring confidentiality of the information provider on elder abuse. Police personnel should be trained in dealing with and taking care of older persons for which police departments should run short courses in collaboration with locally located medical colleges.

Implications for Non-governmental Organizations

There are number of Non-Governmental Organizations (NGOs) working for the welfare of older persons including HelpAge India (Old Age Solutions 2015). The role of NGOs in the welfare of older persons is very important in highly populated countries like India.

Conclusion

Health workers in community and emergency departments are best positioned for early identification of elder mistreatment. It is a challenge as the seniors often hesitate to disclose mistreatments to others including health-care personnel.

Around one in six older persons reported mistreatment. The prevalence of mistreatment of older persons is 15.7%. Neglect was the most common sub-type reported, followed by exploitation, abuse, and abandonment.

Mistreatment of older persons was associated with their psychological distress. Older persons who reported mistreatment had higher psychological distress than those who did not report mistreatment. Older persons who reported abuse, neglect, and exploitation also had higher psychological distress than those who did not report.

Mistreatment of older persons was associated with age of the older persons. Older persons aged 80 and above were more likely to be mistreated than their younger counterparts. Older persons who were living alone or with relatives were more likely to be mistreated than those who were living with their son or daughter. Older persons aged 80 and above and living alone are more prone to mistreatment.

Since the older persons were examined cross-sectionally, it was possible to tell whether the psychological distress existed before mistreatment or was a consequence of the mistreatment.

Under-reporting of mistreatment and exaggeration of social support availability cannot be ruled out due to cultural sensitivity of older persons and their concern about family dignity in society.

There is a need to develop a screening and reporting protocol for primary health-care personnel. Interventions focusing on self-development of older persons, family, community, health-care personnel, and police is essential to tackle this social problem. Nurses can be better utilized for this cause with proper training in prevention of elder mistreatment.

Recommendations: Measures for Prevention and Management of Mistreatment

Risk reduction strategies for older persons, family, and community; family education; and early screening for mistreatment may reduce the risk of elder mistreatment. Early diagnosis, referral, treatment, and counselling services to victim and family may have a positive effect on functional and coping ability of the older person and family. On the other hand the social support of older persons is instrumental in not only preventing mistreatment but also protect the victims from psychological distress and ill health.

The study can be replicated in multiple settings, including rural and urban samples with large sample sizes. Qualitative studies can be conducted to elicit more in-depth information regarding elder abuse and its causes, perception, and perpetrators. Prospective studies to evaluate existing preventive strategies on mistreatment of older persons can be done.

Acknowledgements The authors are thankful to all the participants of the study. There was no conflict of interest. No funds were taken from anyone for the study.

References

Abdel Rahman, Tomader T., and Maha M. El Gaafary. 2011. Elder mistreatment in a rural area in Egypt. *Geriatrics and Gerontology International* 12: 532–537. Accessed 2 December 2015. doi:10.1111/j.1447-0594.2011.00780.x.

Acierno, Ron, Melba A. Hernandez, Ananda B. Amstadter, Heidi S. Resnick, Kenneth Steve, Wendy Muzzy, et al. 2010. Prevalence and correlates of emotional, physical, sexual, and financial abuse and potential neglect in the United States: The National Elder Mistreatment Study. *American Journal of Public Health* 100: 292–297. Accessed 12 December 2015. doi:10.2105/ajph.2009.163089.

Ahmad, M., and M.S. Lachs. 2002. Elder abuse and neglect: What physicians can and should do. *Cleveland Clinic Journal of Medicine* 69 (10): 801–808. Accessed 2 January 2016. doi:10.3949/ccjm.69.10.801.

Alter, Michael J. 2004. *Science of Flexibility*. 3rd ed. Human Kinetics Publishers, Inc.

Amstadter, Ananda B., Kristyn Zajac, Martha Strachan, Melba A. Hernandez, Dean G. Kilpatrick, and Ron Acierno. 2011. Prevalence and correlates of elder mistreatment in South Carolina: The South Carolina elder mistreatment study. *Journal of Interpersonal Violence* 26: 2947–2972. Accessed 8 December 2015. doi:10.1177/0886260510390959.

Atkins, Joanna, Sharon L. Naismith, Georgina M. Luscombe, and Ian B. Hickie. 2013. Psychological distress and quality of life in older persons: Relative contributions of fixed and modifiable risk factors. *BMC Psychiatry* 13. Accessed 5 April 2015. doi:10.1186/1471-244x-13-249.

Birren, J.E., and L.M. Fisher. 1995. Aging and speed of behavior: Possible consequences for psychological functioning. *Annual Review of Psychology* 46: 329–353. Accessed 2 January 2016. doi:10.1146/annurev.ps.46.020195.001553.

Carmona-Torres, Juan Manuel, Pablo Jesús López-Soto, Ana Isabel Coimbra-Roca, Ruth Mary Gálvez-Rioja, Thomas Goergen, et al. 2015. Elder abuse in a developing area in Bolivia. *Journal of Interpersonal Violence.* Accessed 12 December 2015. doi:10.1177/0886260515608803.

Disabled Persons in India, National Sample Survey Organization 58th Round. 2003. Ministry of Statistics and Programme Implementation, Government of India. Accessed 10 November 2015.

Dong, Xinqi. 2009. Elder self-neglect and abuse and mortality risk in a community-dwelling population. *Journal of the American Medical Association* 302: 517. Accessed 10 December 2015. doi:10.1001/jama.2009.1109.

Dong, Xinqi, Melissa A. Simon, Martin Gorbien, Jeffrey Percak, and Robyn Golden. 2007. Loneliness in older Chinese adults: A risk factor for elder mistreatment. *Journal of the American Geriatrics Society* 55: 1831–1835. Accessed 11 December 2015. doi:10.1111/j.1532-5415.2007.01429.x.

Dongre, Amolr, and Pradeepr Deshmukh. 2012. Social determinants of quality of elderly life in a rural setting of India. *Indian Journal of Palliative Care* 18: 181–189. Accessed 15 January 2015. doi:10.4103/0973-1075.105688.

Elder Abuse. 2015. World Health Organization. Accessed 2 November 2015. http://www.who.int/mediacentre/factsheets/fs357/en/.

Elder Abuse. 2016. World Health Organization. Accessed 10 January 2016. http://www.who.int/ageing/projects/elder_abuse/en/.

Elder Abuse and Neglect: In Search of Solutions. 2016. Accessed 10 January 2016. http://www.apa.org/pi/aging/resources/guides/elder-abuse.aspx.

Fulmer, Terry, Rachel F. Rodgers, and Allison Pelger. 2014. Verbal mistreatment of the elderly. *Journal of Elder Abuse and Neglect* 26: 351–364. Accessed 2 December 2015. doi:10.1080/08946566.2013.801817.

Gallegos-Carrillo, Katia, Carmen García-Peña, Carlos Alberto Durán-Muñoz, Yvonne N. Flores, and Jorge Salmerón. 2009. Relationship between social support and the physical and mental wellbeing. *Revista de Investigación Clínica* 61: 383–391. Accessed 2 January 2015. http://www.medigraphic.com/pdfs/revinvcli/nn-2009/nn095f.pdf.

Gupta, Rashmi, and Anoshua Chaudhuri. 2008. Elder abuse in a cross-cultural context: Assessment, policy and practice. *Indian Journal of Gerontology* 22(3&4): 373–393. Accessed 5 October 2015. http://www.gerontologyindia.com/pdf/vol-22-3-4.pdf.

HelpAge India. 2015. A Report on Elder Abuse and Crime in India. Accessed 12 November 2015. https://www.helpageindia.org/pdf/surveysnreports/elderabuseindia2011.pdf.

HelpAge India. 2016. Elder Abuse in India. Accessed 5 January 2016. https://www.helpageindia.org/pdf/surveysnreports/elderabuseindia2012.pdf.

Joseph, Nitin, Maria Nelliyanil, Srirakshar Nayak, Vyom Agarwal, Arjun Kumar, Harsh Yadav, Gourav Ramuka, and Kshirabdhitanaya Mohapatra. 2015. Assessment of morbidity pattern, quality of life and awareness of government facilities among elderly population in South India. *Journal of Family Medicine and Primary Care* 4: 405–410. Accessed 8 December 2015. doi:10.4103/2249-4863.161339.

Kleinschmidt, Kurt C. 1997. Elder abuse: A review. *Annals of Emergency Medicine* 30(4): 463–472. Accessed 5 December 2015. doi:10.1016/s0196-0644(97)70006-4.

Kruger, Maj Robert M., and Capt Christopher H. Moon. 1999. Can you spot the signs of elder mistreatment? *Postgraduate Medicine* 106(2): 169–183. Accessed 5 December 2015. doi:10.3810/pgm.1999.08.657.

Kumar, P.N. Suresh, P.K. Anish, and Biju George. 2015. Risk factors for suicide in elderly in comparison to younger age groups. *Indian Journal of Psychiatry* 57: 249–254. Accessed 15 December 2015. doi:10.4103/0019-5545.166614.

Kurre, S. 2004. Elder abuse. *Australian Family Physician* 33(10): 807–812. Accessed 5 December 2015.

Kurz, Rosemarie. 2015. Who is considered an old person? Speech, The EURAG Congress, Spain, Leon, November 15, 2015. http://generationen.oehunigraz.at/files/2012/07/Who-is-considered-an-old-person.pdf.

Loughlin, Anne O., and Joseph Duggan. 1998. Abuse, neglect and mistreatment of older people: An exploratory study. *National Council on Ageing and Older People* 52: 1–122.

Lowenstein, Ariela, Zvi Eisikovits, Tova Band-Winterstein, and Guy Enosh. 2009. Is elder abuse and neglect a social phenomenon? Data from the first national prevalence survey in Israel. *Journal of Elder Abuse and Neglect* 21: 253–277. Accessed 2 January 2016. doi:10.1080/08946560902997629.

Luo, Y., and L. J. Waite. 2011. Mistreatment and psychological well-being among older adults: Exploring the role of psychosocial resources and deficits. *The Journals of Gerontology Series B: Psychological Sciences and Social Sciences* 66B: 217–229. Accessed 10 December 2015. doi:10.1093/geronb/gbq096.

Martins, Rosa, Maria João Neto, Ana Andrade, and Carlos Albuquerque. 2014. Abuse and maltreatment in the elderly. *Atención Primaria* 46: 206–209. Accessed 10 December 2015. doi:10.1016/s0212-6567(14)70093-9.

Mental Health and Older Adults. 2016. World Health Organization. Accessed 2 January 2016. http://www.who.int/mediacentre/factsheets/fs381/en/.

Missing Voices: Views of Older Persons on Elder Abuse. 2016. World Health Organization. Accessed 12 January 2016. http://www.who.int/ageing/publications/missing_voices/en/.

Nainital District: Census 2011 Data. Accessed 5 January 2016. http://www.census2011.co.in/census/district/584-nainital.htm.

National Policy on Older Persons. 1999. Ministry of Social Justice and Empowerment, Government of India. Accessed 12 December 2015. http://socialjustice.nic.in/hindi/pdf/npopcomplete.pdf.

Naughton, C., J. Drennan, I. Lyons, A. Lafferty, M. Treacy, A. Phelan, A. O'loughlin, and L. Delaney. 2011. Elder abuse and neglect in Ireland: Results from a national prevalence survey. *Age and Ageing* 41: 98–103. Accessed 20 November 2015. doi:10.1093/ageing/afr107.

Nisha, Catherin, Steve Manjaly, Pretesh Kiran, Betsy Mathew, and Arvind Kasturi. 2015. Study on elder abuse and neglect among patients in a medical college hospital, Bangalore, India. *Journal of Elder Abuse and Neglect* 28: 34–40. Accessed 10 January 2015. doi:10.1080/08946566.2015.1113599.

Old Age Solutions. 2015. NGOs and Voluntary Organizations. Accessed 10 December 2015. http://www.oldagesolutions.org/facilities/NGOVOL.aspx.

Older People's Health Issues. 2016. Merck Manuals Consumer Version. Accessed 5 January 2016. http://www.merckmanuals.com/home/older-people%E2%80%99s-health-issues.

Oliveira, Amanda Maria Souza De, and Larissa Chaves Pedreira. 2012. Being elderly with functional dependence and their family caregivers. *Acta Paulista de Enfermagem* 25: 143–149. Accessed 5 December 2015. doi:10.1590/s0103-21002012000800022.

Problems of Elderly. 2015. Help-age India. Accessed 5 January 2015. http://www.helpageindiaprogramme.org/Elderly%20Issues/problems_of_the_elderly/index.html.

Protecting Elders Against Neglect and Abandonment. 2015. Accessed 10 November 2015. http://www.swd.gov.hk/doc/family/elderly-NE&AB.pdf.

Rajan, S. Irudaya. 2006. Population ageing and health in India. *The Centre for Enquiry into Health and Allied Themes*. Accessed 5 December 2015. http://www.cehat.org/humanrights/rajan.pdf.

Saikia, Ankumoni, Neelakshi Mahanta, Ajaya Mahanta, Ashokjyoti Deka, and Arupjyoti Kakati. 2015. Prevalence and risk factors of abuse among community dwelling elderly of Guwahati City, Assam. *Indian Journal of Community Medicine* 40: 279–281. Accessed 14 December 2015. doi:10.4103/0970-0218.164406.

Salokangas, Raimo K.R., and Matti Joukamaa. 1999. Physical and mental health changes in retirement age. *Psychotherapy and Psychosomatics* 55(2–4): 100–107. Accessed 3 January 2016. doi:10.1159/000288415.

Sampath Kumar, S., S. Girijakumari, and V. Brindha. 2015. Understanding elderly abuse—A special reference to elderly in an urban slum of Coimbatore, Tamil Nadu. *International Journal of Scientific and Research Publications* 5: 1–3. Accessed 10 December 2015. http://www.ijsrp.org/research-paper-0615/ijsrp-p4273.pdf.

Satheesh, N.V., Anuj Nautiyal, and Madhav Abhijeet Ojha. 2015. Prevalence of depression among geriatric people in Dehradun City of Uttarakhand, India. *Journal of Depression and Anxiety* 04. Accessed 9 December 2015. doi:10.4172/2167-1044.1000208.

Sharoni, Siti Khuzaimah Ahmad, Emad Adel Shdaifat, Hayati Adilin Mohd Abd Majid, Norhafizatul Akma Shohor, Fazimah Ahmad, and Zalina Zakaria. 2015. Social support and self-care activities among the elderly patients with diabetes in Kelantan. *Malaysian Family Physician* 10: 34–43. Accessed 10 December 2015. https://www.ncbi.nlm.nih.gov/pmc/articles/PMC4567891/.

Shivakumar, Prafulla, Shilpa Sadanand, N. Girish, Mariamma Philip, and Mathew Varghese. 2015. Identifying psychological distress in elderly seeking health care. *Indian Journal of Public Health* 59: 18–23. Accessed 3 December 2015. doi:10.4103/0019-557x.152849.

Skirbekk, Vegard, and K.S. James. 2014. Abuse against elderly in India—The role of education. *BMC Public Health* 14: 1. Accessed 15 December 2015. doi:10.1186/1471-2458-14-336.

Srinivasan, Chokkanathan. 2015. Elder mistreatment and health status of rural older adults. *Journal of Interpersonal Violence* 30(18): 3267–3282. Accessed 5 January 2016. doi:10.1177/0886260514555014.

Srinivasan, Chokkanathan, and Alex E.Y. Lee 2005. Elder mistreatment in urban India: A community based study. *J Elder Abuse & Neglect* 17: 45–61.

Strasser, Sheryl, Megan Smith, Scott Weaver, Shimin Zheng, and Yan Cao. 2013. Screening for elder mistreatment among older adults seeking legal assistance services. *Western Journal of Emergency Medicine* 14: 309–315. Accessed 12 January 2015. doi:10.5811/westjem.2013.2.15640.

The Council on Scientific Affairs, AMA, Chicago. 1987. Elder abuse and neglect. *The Journal of the American Medical Association* 257(7): 966–971. Accessed 5 December 2015. doi:10.1001/jama.1987.03390070086030.

Uttarakhand Population Census Data. 2011. Accessed 5 January 2016. http://www.census2011.co.in/census/state/uttarakhand.html.

Wang, Jing-Jy, Jong-Ni Lin, and Feng-Ping Lee. 2006. Psychologically abusive behavior by those caring for the elderly in a domestic context. *Geriatric Nursing* 27: 284–291. Accessed 15 December 2014. doi:10.1016/j.gerinurse.2006.08.016.

Wu, Li, Hui Chen, Yang Hu, Huiyun Xiang, Xiang Yu, Tao Zhang, Zhongqiang Cao, and Youjie Wang. 2012. Prevalence and associated factors of elder mistreatment in a rural community in People's Republic of China: A cross-sectional study. *PLoS ONE* 7. Accessed 22 December 2015. doi:10.1371/journal.pone.0033857.

Wu, L., Y. Hu, X. Yu, T. Zhang, Z.Q. Cao, and Y.J. Wang. 2015. Association between social support and mistreatment on elderly in the rural community of Hubei province (Abstract). *Zhonghua Liu Xing Bing Xue Za Zhi* 32: 978–982. Accessed 5 November 2015. http://www.ncbi.nlm.nih.gov/pubmed/22333078.

Chapter 9
Are Elderly Widows More Vulnerable to Abuse and Violence? Findings from Jharkhand, India

Sangeeta Kumari Gupta and T.V. Sekher

Feminization of Ageing in India

The ageing of a population is an outcome of an increase in the proportion of the aged *vis-a-vis* a decrease in the proportion of the young. Improved life expectancy has given rise to an increase in the number of persons aged 60 years and above. In 1950, the proportion of elder persons in the world was around 8% and it has increased to 10% in 2000 and projected to reach 21% by 2050. It is estimated that the elderly population will reach almost 1.2 billion by 2025 and 2 billion by 2050 (United Nations 2002).

The India's aged population is currently the second largest in the world. The share of India's elderly population is projected to climb to 19% by 2050, leading to 323 million old people. This will raise a variety of social, economic and health care challenges. More than 50% of elder women in India live without a spouse as compared to only 15% of elder men. The increasing proportion of elderly women, especially widows is a characteristic feature of India. The proportion of elderly (60 and above years) in India's population is 8.6% and likely to increase significantly in coming decades (Census of India 2011). India's aged population suffers from gradual 'feminization' and serious gender disparities which is evident from the growth pattern of the elderly females (Alam 2006), which is a matter of serious concern. Another important feature in the Indian context is the central role played

Some of the arguments and data used in this chapter are similar to those of a recently published paper by the same authors: Gupta and Sekher (2017).

S. Kumari Gupta (✉) · T.V. Sekher
International Institute for Population Sciences (IIPS), Mumbai, India
e-mail: sangeetasargam@gmail.com

T.V. Sekher
e-mail: tvsekher@gmail.com

Mala Kapur Shankardass and S. Irudaya Rajan (eds.), *Abuse and Neglect of the Elderly in India*, https://doi.org/10.1007/978-981-10-6116-5_9

Table 9.1 India's aged population by age, sex and marital status, 2011

Aged	Sex	Marital status (proportion)			
		NM	M	W	D/S
60+ years	Males	2.9	82.1	14.6	0.4
	Females	2.0	49.6	47.8	0.5
60–64	Males	2.1	88.8	8.7	0.4
	Females	1.6	63.0	34.7	0.8
65–69	Males	2.8	84.9	11.9	0.4
	Females	2.0	54.4	43.0	0.6
70–79	Males	2.2	78.3	19.1	0.4
	Females	1.6	37.1	60.8	0.4
80+ years	Males	7.5	62.0	30.1	0.4
	Females	4.5	26.1	69.0	0.4

Source Census of India (2011)
NM never married, *M* married, *W* widowed, *D/S* divorced/separated

by the family and the limited role of the state in acting as a safety net for the elderly in terms of various forms of support such as co-residence, economic support and care-giving (Rajan and Balagopal 2017).

The proportion of widows in the female population rises sharply with age, it is 48% among women aged 60 and above, and it reaches to 61% among women aged 70 and above (Table 9.1). There are three reasons for this unusual gender disparity in widowhood in India—the longer life span of women than that of men, the cultural practice of Indian men prefer to marry women considerably younger than themselves, and the higher rate of re-marriage among the widowed men compared to that of widowed women.

In Indian context, widows are the unfortunate victims and for centuries, they have been ill-treated by families and society (Pandya and Priti 2006). Regardless of the increasing education and economic development happening in India, the status of women within the traditional households continues to be the same. Despite improvements in female education and continuing attempts for social reforms, the position of elderly widows are not very encouraging. This is more so in the rural areas. For large number of women, '*Widowhood is more than the loss of a husband —it may mean the loss of a separate identity*' (UNFPA 1998: 42). They are mostly dependents on their family throughout their life. More than half of the elderly were economically dependent on others for their day-to-day requirements, as the widowhood and dependency increases with advancing age, the situation is worse for the elderly females (Kumari and Sekher 2010; Gupta and Sekher 2017).

Social networks and family dynamics are central to the well-being of elderly, with positive and negative aspects (Berkman et al. 2012). Many times, a widow faces difficulty in adjusting to her new status because the social norms and taboos place her in most disadvantaged position and restrict her social interactions. A widow has to perform many rituals. She has to follow lot many restrictions leading to social isolation and social inferiority, resulting in a feeling of dejection and depression. Widows in all communities suffer from limited freedom to participate in religious and social ceremonies. They carry the stigma of inauspicious and hence have to keep away from social functions. '*A widow is supposed to be disfigured; her ornaments are removed, her heads shaved, and she has to dress white sari. After that she is expected to give up eating 'hot foods, to avoid auspicious occasions because she is considered inauspicious and to remain celibate, devout and loyal to her husband*' (Chen 1998: 26), but no such stigma or restrictions imposed on widowers.

Indian widow tends to be a highly marginalized person. Beyond the economic difficulty, they often face social and cultural problems. She typically receives very little support from persons other than her own children and even when she lives with her adult sons, she remains highly vulnerable to neglect (Sivagami and Thamilrasan 2004; Gupta 2017). Sebastian and Sekher (2011) and Sebastian (2011) found that among Indian families, for a variety of reasons and circumstances, elderly widows are more vulnerable to abuse and neglect than others.

Elderly widows more vulnerable to loneliness because of their greater distance from neighbours and services. Visits to family members make little difference in the older person's feeling of loneliness or life satisfaction (Gerg 1976). Level of interaction with the family members, friends and neighbours can affect their mental health. Further it was observed that poor health, less physical mobility, less satisfaction, and frequency of interaction with family and neighbours are often associated with experiencing loneliness among the elderly (Carolyn 1981). There is no statistics, which indicate the extent and depth of the vulnerability among India's elderly widows (Gupta and Sankar 2002). However, the incidence and prevalence of elder abuse is on the rise and gaining visibility in many new forms (Shankardass 2010).

The National Policy for Older Persons of the Government of India (1999) assures elderly that their concerns are national concerns and they will not live unprotected, ignored or marginalized. However, the needs and requirements of widows have not given priority in the welfare policies. Widowhood is a turning point in a woman's life, involving new social adjustments for the family as well as for her. These women constitute a special group in our population because of the deprivation, vulnerability and low social status. Therefore, the present study try to understand the ill-treatment and neglect experienced by elderly widows within their own households and immediate surroundings.

Methods, Data and Field Work

Both quantitative and qualitative data have been used in this study. In quantitative component, the survey data collected from selected villages during 2010 has been analyzed. The case studies and key informant interviews were used to explore issues related to vulnerability, experience of ill-treatment/abuse, and the extent and nature of deprivation, loneliness, etc.

Selection of Study Areas: For the study purpose, Jharkhand state was purposively selected. East Singhbhum district was selected for the study, since this district was having higher proportion of elderly widows. Two blocks were then randomly selected, from the district. It was targeted to personally interview 300 elderly widows. In total, 12 villages from the two blocks were visited, in order to complete the interviews of 300 elderly widows.

Preliminary visits to the study area before initiating the household survey helped in establishing good rapport with villagers. From the first block, 7 villages were covered in the survey. In the second block, 5 villages were identified to collect information from elderly widows. With the help of *Sarpanch* (village head), *Anganwadi* (ICDS) workers and other functionaries from the villages, the households with eligible respondents were identified and personally interviewed using a structured schedule. Care was taken to ensure privacy as far as possible during interviews/data collection.

Profile of the Surveyed Households and Elderly Widows

Households Characteristics of the Surveyed Population

Eighty percent of surveyed household were Hindus, only 7% were Muslims, 13% belonged to the *Sarna*[1] religion (Table 9.2). 40% of households were scheduled tribes, followed by other backward castes (37%). The average household size was 4.7. 96% of surveyed households had one elderly person and around 3% of households had two elderly persons. More than half of the households were female headed. 55% of the households owned no land. Little less than one-fourth (24%) of the households were having large holdings (land size greater than 3.2 acres). Further, to find out the income from the land, a question was asked—'*How much average yearly income you get from your land?*' 23% of the households owning land reported no income from the land. However, 37% of the households reported some income from the land and the average is Rs. 3500 in a year.

[1]According to Troisi (2000), Sarna is the Munda word for 'Sacred Grove' while Dhorom is the Oriya word meaning 'religion'. Sarna involves belief in a great spirit called the Sing Bonga. Santhal belief holds the world to be inhabited by numerous spiritual beings of different kinds. Santhals consider themselves as living and doing everything in close association with these spirits. Rituals are performed under groves of Sal trees called Jaher (or sacred grove), where Bonga is believed to appear or express himself. Often Jaher are found in the forests.

Table 9.2 Distribution of elderly widows by selected socio-economic and demographic characteristics

Background characteristics	Percent	Number
Age-group		
Young-old (60–69 years)	64.3	193
Old-old (70–79 years)	27.0	81
Oldest-old (80 and above years)	8.7	26
Religion		
Hindu	80.3	241
Muslim	7.0	21
Sarna	12.7	38
Caste		
Scheduled Castes (SCs)	14.3	43
Scheduled Tribes (STs)	40.0	120
Other Backward Castes (OBCs)	37.3	112
Others	8.3	25
Literacy		
Literate	9.7	29
Non-literate	90.3	271
Number of sons		
No son	21.3	64
One son	28.3	85
Two sons	27.0	81
Three or more sons	23.3	70
Number of daughters		
No daughter	21.7	65
One daughter	26.7	80
Two daughters	24.0	72
Three or more daughters	27.6	83
Current work status		
Working	29.7	87
Not working	70.3	213
Occupational status[a]		
Non-agricultural labour	47.1	41
Agricultural labour	29.9	26
Self-employed/petty business	20.7	18
Salaried	2.3	2
Total number of elderly widows	100.0	300

[a]Among the respondents working at the time of survey

Housing Characteristics

Majority of households (94%) owned the structure in which they resided. In case of the housing type, it was observed that about three-fifths of households (60%) lived in *kachcha* houses (constructed from mud or other low quality materials), 16% lived in *semi-pucca* houses and 24% lived in *pucca* houses. 28% of the residential structures consist of single room however, more than half of the residential structures contained 2–3 rooms. Less than half (40%) of the households had separate kitchen for cooking, around one-third of the households cook inside their living rooms and one-third of the households cook in open areas. 89% of all surveyed households were using dry leaf/wood/coal/charcoal/dung cakes as main source of cooking fuel. The use of LPG was very less (around 8%). 70% of the households visited do not have access to basic sanitation (toilet facility). Around one-fourth of households are using kerosene as a main source of lighting. 57% households reported that their main source of drinking water is public piped water/hand pump/covered well. Only 16% of households use own piped water, water from own hand pump or covered well.

Households/Elderly Benefited from Government Schemes

To understand whether the households benefited from government programmes, information was collected. Nearly half of the households were having below poverty line (BPL) cards and using this card, they were receiving food grains and kerosene. 17% of the households were getting wheat under the provision of BPL card. Around half of the elderly interviewed were not receiving any pension. Among those receiving, 74% got widow pension and 19% getting old-age pension. The amount of widow pension and old age pension is Rs. 200 in a month at the time of survey (now increased to Rs. 600). It was observed that all elderly widows were keen to receive pension. When enquired about the reasons for not availing the pension, they cited various reasons—'we do not know how to apply, officers are not very helpful, we are not eligible, etc.'. Few households benefited from the *Indira Aawas Yojana* (free housing scheme of the government).

In order to examine current economic status of the households, a question was asked-'*whether the household had borrowed money for any purpose during the last one year? If yes, for what purpose they borrowed and the source of borrowing*'? 22% of the households had borrowed money during the last year. The most important reasons mentioned were 'treating for health problems' (38%), followed by 'meeting the household consumption' (23%). 12% of households have borrowed money in order to 'meet marriage related expenses'. Major source of borrowing was neighbours (39%) and relatives (23%). Only 15% have taken loans from banks and 14% from local money lenders. If they avail loan from relatives and neighbours, they need not to pay interest. Money lenders charge higher interest than banks. 30% of the households have indebtedness of around Rs. 700–Rs. 2000, whereas slightly less than one-fifths of the households have indebtedness of more than Rs. 8500.

Profile of the Elderly Widows Interviewed

Among the respondents, 64% of the elderly widows are in the young-old age group (60–69 years) followed by old-old age group (70–79 years) and oldest-old (80 and above years) age groups (27 and 9% respectively) (Table 9.2). As expected, the literacy rate was very low among the surveyed elderly widows. Around one-third among them were working at the time of survey. Among those working, 30% were engaged in agricultural work and 47% were non-agricultural laborers, 21% were self-employed, and only 2% were salaried employees. More than half of the respondents (67%) were staying in nuclear families. About 13% of elderly widows interviewed in this study were staying alone or forced to live alone.

Living Arrangements of Elderly Widows

Family is the most important social unit and the major source of care and support to the aged, especially for widows. Living arrangements of the elderly make a significant difference to a variety of factors that may directly or indirectly contribute to the quality of their life and welfare. It also provides certain indications regarding the level of actual support available to the elderly. India is characterized by its traditional ways of living where several generations live together within the same household. However, the traditional ways of the extended family members living together are now gradually breaking down. The elderly have been the worst sufferers of the changing family structure in India. The living arrangements of the elderly have been severely affected by changes in the multigenerational familial co-residence and compelling them to live by themselves or alone, if widowed.

A question was posed to the respondents about their preferences in living arrangements. Among the elderly widows who were currently living alone, 67% prefer to live alone and little less than one-fourth (23%) want to live with their

Table 9.3 % distribution of elderly widows according to their preferred living arrangements and actual living arrangements

Actual living arrangements	Preferred living arrangements				
	Living alone	With unmarried children	With married son/s	With married daughter/s	With other relatives
Living alone	66.7	0.0	23.1	2.6	7.7
Living with unmarried children	9.1	81.8	9.1	0.0	0.0
Living with married son/s	7.8	0.6	91.0	0.6	0.0
Living with married daughter/s	0.0	0.0	5.6	94.4	0.0
Living with other relatives	15.2	0.0	9.1	0.0	75.8

married sons. Among the respondents who are currently residing with their married sons, majority prefers to live with them. However, 8% among them prefer to live alone. Elderly widows who were currently living with their married daughters, 6% prefer to live with their married sons. Among those who were currently living with other relatives, 15% prefer to live alone (Table 9.3).

Perception of Status Among Elderly Widows

Good number of elderly widows (42%) interviewed felt that their status is low in the village. Interestingly, only one-out of-seven really believed that widows have good status in the community. When probed about the reasons attributed for having low status in the village, majority consider poor economic conditions. One-fourth of elderly widows believed that social factors are responsible for the persisting low status. To understand their status within own households, a question was posed to every respondent—'*In your opinion, what is your status within your own family?*' The response options were—*low, medium, or high.* Nearly half of them (49%) have opined that they consider to have only medium status, whereas around one-third felt that their status is low. However, more than half (59.8%) of the elderly widows were satisfied with their present status and 40% were not satisfied with. Majority (59.7%) felt that their status has changed considerably after they became widow. Nearly half of the interviewed women mentioned that they experienced a lack of respect from own family after widowhood. One-fourth of them opined that they were not consulted at all in any family decision-making. Around 12% of the elderly widows reported that they have to take permission from their household members on important matters, which is not a pleasant experience. What is revealing is the fact that one-in-ten elderly women were not involved in family functions after becoming widow (Table 9.4).

In this study, an attempt was made to examine whether elderly widows are satisfied with their present status in the family. As expected majority of elderly widows, who perceive their status is low, are not satisfied. However those who felt to have better status in the family, around 90% of them expressed satisfaction in the present situation. Another important dimension to be noted here is that around half of the economically dependent widows stated that they are not at all satisfied with their present status in their own families (Table 9.5).

In many instances, the state of widowhood is exacerbated by the conditions of poverty (United Nations 2001). We observed that 81% of elderly widows had faced many problems immediately after becoming widow. More than half of the elderly widows have faced economic difficulties. Around 35% were faced with childcare problems. Around 14% reported to have experienced 'social stigma'. In the study villages, women were blamed for the death of their husbands. A 65 years old tribal widow revealed—'*After my husband's death, people said that I am a 'witch' and I only killed my husband*'. She further stated—'*My neighbour one day warned me - if you want to be alive, then do not show your face to anyone outside*' (Case study-1,

Table 9.4 Elderly widows' self-perception about their status within and outside family

Elderly widows perceived status within and outside their family	Percent	Number
Status of widows in the village		
Low	42.0	126
Medium	51.0	153
High	7.0	21
Factors responsible for the low status of widows		
Economic factors	49.5	149
Social factors	23.7	71
Family factors	21.0	63
Religious factors	5.7	17
Elderly widow's status within own family		
Low	32.3	97
Medium	49.3	148
High	18.3	55
Are you satisfied with your status within your family?		
Yes	59.7	179
No	40.3	121
Whether your status in the family has changed after becoming widow?		
Yes	56.7	170
No	43.3	130
If yes, In which aspects, your status has changed?[a]		
Lack of respect	47.1	80
Opinion not sought	25.3	43
Need permission for everything	11.8	20
Not involved in family functions	9.4	16
Other reasons[b]	6.4	11

Source This table is based on the primary survey conducted by authors in the selected villages of East Singhbhum district of Jharkhand

[a]Among those who said their status has changed after becoming a widow (N = 170)

[b]Other reasons include health related problems, economic problems, loneliness, etc.

Table 9.5 Elderly widows' satisfaction and level of economic dependency (in %age)

Perceived status of elderly widows within family	Satisfied with status within family		
	Yes	No	Number
Low	18.6	81.4	97
Average	75.0	25.0	148
High	90.9	9.1	55
Level of economic dependency			
Fully dependent	68.6	31.4	121
Partially dependent	64.2	35.8	53
Not dependent	49.2	50.8	126

Table 9.6 Treatment of elderly widows by husband's family, before and after widowhood

Treatment of elderly widows by husband's family	Before widowhood	After widowhood	Paired *t* value
Bad	22.3	45.3	8.938*** (0.325–0.508)
Indifferent	28.0	23.7	
Good	49.7	31.0	

Figures in parentheses is 95% confidence interval
Significance level given as *** $P < 0.01$

see Sect. "Annexure"). In many villages of these backward region, widowhood is still a curse for women. They were humiliated and abused by people, including own family members, for no reason.

After the death of only earning member in the family, many widows find it extremely difficult to manage the household all alone. Widowhood is embedded with emotional shocks, a sudden and drastic change in social status, and often an economic crisis for many (Bhatt 2006). The help from community and family members were a great source of support during this difficult and testing times. But, nearly half of the elderly widows interviewed did not get any help immediately after their husband's demise. In this context, it is important to examine the widow's continued relation/contact with her husband's family. A question was asked to the respondents—'*How is your relationship now with members of your husband's family?*' One third (30%) of the elderly widows stated that they do not have any contact with their husband's family now. This is a good indication of their gradual isolation from own relatives.

Did elderly widows receive the same kind of respect and care from husband's family before and after widowhood? In order to capture this, a question was posed to respondents—'*Before and after the death of your husband, were you treated with same respect by your husband's family?*' More than half of elderly widows stated that husband's family did not treat them with same respect after widowhood (Table 9.6).

There is considerable increase in the percentage of elderly widows who were treated badly by the family members of husband. 22% of the elderly widows were treated badly by their husband's family before widowhood. But after the widowhood, this has gone up to 45%, illustrating the changed attitude and behaviour of husband's family. In the survey, information was gathered about the participation of elderly widows in social and religious activities. Half of the respondents stated that they usually participate in family-get-together. They do participate in social gatherings like marriage, funeral or any other functions. Some of the elderly widows revealed that some people never invite them to social or religious functions.

While interviewing, one elderly widow said—'*I attend marriages in our village. But when new bride comes to the house, I am not allowed to go there and see her*'. She further mentioned—'*Even if new bride comes to our house, I cannot welcome her and perform any ritual*'. After widowhood, there is a significant decline in the participation in social activities. This reflected greater isolation and a lack of community identity for the elderly widows. Paired *t* test infers there is a significant difference among the elderly widows for attending the social activities before and after the widowhood (Table 9.6).

Mistreatment or Abuse Experienced by Elderly Widows

'*Life has become a hell for me. Who will take care of a woman when she is widowed?*' (said a widow, aged 64 years). Many elderly widows are exposed to abuse and mistreatment within their own families. Sebastian and Sekher (2011) and Kumari and Sekher (2012) found that in Indian households, the female elderly, particularly aged widows, are more susceptible to abuse and neglect for obvious reasons.

American Psychological Association (2004) defines elder abuse as—'It is the infliction of physical, emotional, or psychological harm on an older adult. It also can take the form of financial exploitation or intentional or unintentional neglect of an older adult by the caregiver'. To examine the abuse or mistreatment experienced by the elderly widows, few questions were asked—'*Do you try to avoid some members of this family?*' followed by another question—'*Are you afraid of anybody with whom you are currently staying?*' These questions were canvassed to the respondents by ensuring that no one else was present at the time of interview (Fig. 9.1).

Seven percent of elderly widows interviewed were afraid of someone in their own family. Being widowed and aged, they are more vulnerable to abuse and neglect. Around 5% of elderly widows reported that they were physically harmed by someone in the household.

An elderly widow (83 years) narrated her experience—'*My daughter-in-law always quarrel with me. My son and daughter-in-law are not taking care of me. My neighbours help me in doing my work*'. In her own words—'*Isi din ko dekhane ke liye itana kast karke bete to aadmee banaye, par mujhe kya pata that ki beta aisa ho*

Fig. 9.1 Mistreatment experienced by elderly widows within the family (in %age)

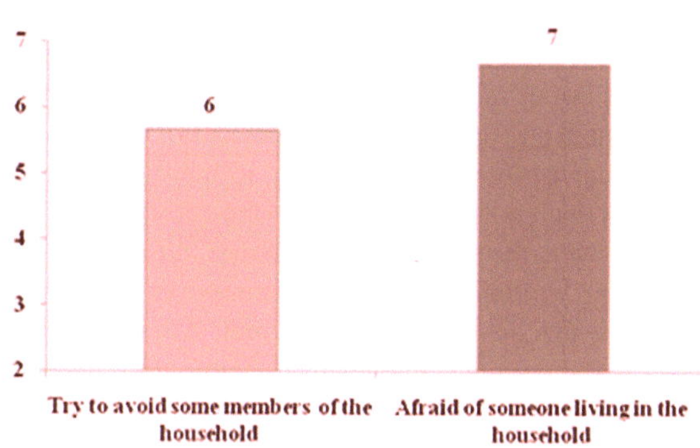

Fig. 9.2 Neglect experienced by elderly widows (in %age)

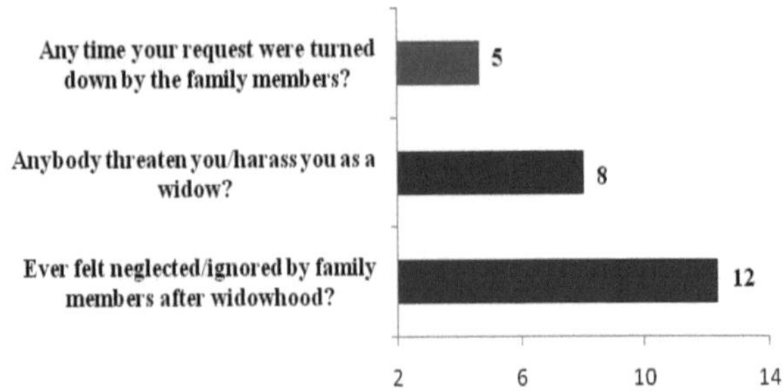

Table 9.7 Depression among elderly widows

Level of depression	Percent	Frequency
No depression	22.3	67
Mild depression	46.7	140
Severe depression	31.0	93
Cronbach's alpha value	0.6516	

jayega', ('*I have taken lots of pain to bring up my son. For all my hard work of all these years, now he treats me like this!*').

Many elderly widows are considered as a burden by family members. Young children also do not like elderly interfering in their personal matters. Figure 9.2 shows that 12% of elderly widows from the study villages felt neglected or ignored by their own family members. Around 8% of elderly widows revealed that someone is threatening them and 5% have said that family members mostly ignored their requests for help.

Depression and Loneliness Among Elderly Widows

Loss of a spouse at any age has very significant impact on the psychological status of an individual. To measure the level of depression or psychological condition of elderly widows, Geriatric Depression Scale (GDS) was used in this study. GDS consists of 30-item questionnaire in which participants are asked to respond by answering 'yes' or 'no' in reference to how they feel over the past one month. In order to adapt the scale to the local situation, certain questions were excluded from the scale as it may not be relevant in Indian context. Certain questions were modified as per the requirement and the scale was finalized. For the purpose of the present study, 24-items have been considered in the scale. Reliability test has been conducted. One point is given for each positive response to the question and by totaling the points counted for each answer, the score of 0–8 shows '*no depression*' or '*normal*'; score of 9–17 indicating '*mild depression*' and score over 18 shows '*severe depression*'. The time frame for this scale is last one month. After checking the reliability, the composite index was computed. The levels of depression, measured based by administering this scale among the elderly widows, is presented in Table 9.7.

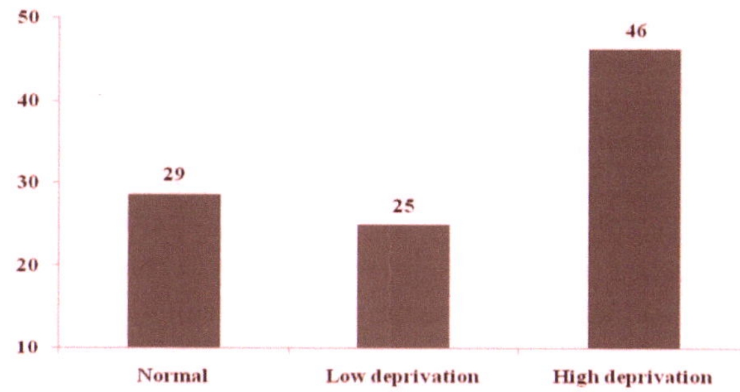

Fig. 9.3 Elderly widows feel emotionally deprived by their family (%)

Around 46% of widows were having mild depression. Among the oldest-old widows, around 35% were severely depressed. Chi-square test was performed to show the association between the background variables and level of depression among the elderly widows. Number of sons, current work status, type of family, economic dependency, and receiving pension emerged as significant factors at 10% level of significance. Current health status was significant at 1% level (not shown in the table here).

Further to study the emotional deprivation among elderly widows, a set of statements were asked to the respondents. The statements were framed in a simple way that it could be answered either as '*yes*' or '*no*'. The statements are—'*The members of this family often go out of their way to help you? Do you think that the members of this family are really interested in you and your problems? The members of this family get along each other as you would like? This family gets together for holiday/festival celebrations? Members of this family respect your rights? You feel loved and cared for by this family? Are you troubled because members of this family differ with you on beliefs and values?*' To capture the extent of emotional deprivation among elderly widows, a composite index was computed using the seven statements canvassed. After checking for the reliability with alpha values (0.8649), the composite index on the level of deprivation has been finalized. The score ranging from 0 to 7 indicates the various levels of emotional deprivation among them. Based on total scores and the mean value obtained, that has been categorized into three—'*normal*', '*low*' (score ranging from 1 to 2) and '*high*' deprivation (score 3 and above) (Fig. 9.3).

According to this composite index, 46% of elderly widows believed to face high level of deprivation at emotional level. More than half of the oldest-old widows have high deprivation on emotional level, may be due to reasons such as—forced to stay at home most of the time, nobody there to interact with them, etc. As expected, higher emotional deprivation was observed among the elderly widows who are living alone. Further, it was also observed that the economic dependency and emotional deprivation are significantly co-related. Loneliness is a common problem among widows and is a kind of situation, which emerges due to the setbacks in the family cohesiveness. It also indicates the disorganization in personality, among elderly in particular (Yadava 2011). A question was asked in the survey—'*How frequently do you feel lonely, even when you are with your family?*' The responses to this question were in three categories—never, sometimes and often. One-fourth of young-old widows and 35% of oldest-old widows often feel loneliness. According to the type of family, more than three-fifths (64%) of elderly widows who stay alone often experience loneliness. The type of family is found to be significantly associated with loneliness. 34% of elderly widows living with their unmarried children often feel loneliness. Those who are living with married children may be engaged with their grandchildren to some extent.

Conclusions: Myth and Reality

The widowhood places the women in a disadvantaged position, particularly in their old ages. Elderly widows experience several problems like low status within family as well as in the society, loneliness, economic hardships, depression, feeling of insecurity, lack of emotional care from family, etc. Along with poverty and economic constraints, ill-treatment and neglect from own households makes them doubly vulnerable and depressed.

The study clearly illustrates elderly widows are the victims within their own households and also within the community. Although most elderly widows reported to have good relationship with their neighbours, but not many have someone close and intimate enough to share their problems and worries. Traditionally in Indian households, elderly are given a respectable role in the family decision-making on social matters. But what we found in this study is that in issues like financial matters, elderly were seldom consulted. More than half of the respondents felt that their status in the family has considerably lowered after the death of their husbands. For this, the economic constraints were the main factors to a large extent. Loneliness is a common problem among widows. Women feel lonely more than men and the loss of the spouse accentuates these differences. The death of a spouse means an end to daily conversation and also an end to many shared activities in which the spouse was the driving force. Death of husband can affect the relationship between family members. The traditional norms and values of the Indian society always provided respect and care to the aged. The aged members are normally taken care by the family. Few elderly widows in the study area are silent victims of mistreatment or abuse within the family. In recent times, due to the gradual breaking down of joint family system, large number of old parents are not maintained by their children. Consequently, the elders are now exposed to emotional neglect and lack of physical and financial support. Since no option left, these women silently suffer the neglect and humiliation. High level of illiteracy, lack of remunerative occupation, as well as lack of awareness about legal and economic rights makes them more vulnerable to neglect and abuse. Majority blamed their fate for the pathetic condition.

In brief, the present study found that elderly widows face discrimination both at the household and community levels which varies from economic, physical, to emotional needs. Being aged as well as widowed, they do not have intimate persons with whom they can share their problems. Living in poverty and uncertainties, the mistreatment and neglect by own family members makes their life further miserable. As seen from the qualitative interviews and case studies, most of them silently cope up with this situation. Widowhood is generally stressful for women, not only a loss of long-time companion, but also the weakening of economic power and social networks. It is suggested that massive and intensive campaigns should be done to sensitize people to stop all forms of abuse and discrimination associated with widowhood. The family and multi-generational living arrangements needs to be strengthened as a viable support system for providing love, care, and protection for older persons, particularly to elderly widows.

Annexure

Case Study 1

Mrs. X. is a 65 years old widow belonging to a tribal family. Five years ago, her husband expired as he was suffering from Tuberculosis. She has one son and three daughters. All her children are married. She earns little money by working as a maid in nearby homes. Her husband was unemployed. He used to take away her money for drinking alcohol.

After husband's death, her neighbours started calling her a '*witch*' ('*Daayan*'). Her neighbours do not like her and not even talk to her. She is considered as 'inauspicious'. Villagers believe that seeing her is a bad omen. They are also scared that she can do 'black magic' and hurt people. They want to get rid of her from the village. '*If you want to be alive, then do not show your face to anyone and stay at home*'. No one is ready to listen to her, for them she is a 'witch' in their vicinity. Everyone started blaming her for her husband's death. Because of people's attitude, she could not work outside. She is completely helpless and frustrated. *She asked —'Am I look like Daayan?*' In her own words—'*agar kisi ka pati kam umar main mar gaya to kya wo aurat dayan hoti hai?*' ('*If the husband die at a very young age, how the wife becomes a witch?*').

During the fieldwork in that village, some women warned me—'*you have to be careful here as some of the widows are witches. They know black magic and can harm you*'. I just asked them—'*if she does something bad to me, then what I can do?*' They replied—'*In our village we have one tantric, who can help to get rid of black magic. However, it is better for you to avoid meeting those women*'.

When I went to the house for interviewing her, she was courteous. While talking to me, she opened up her feelings. '*Everyone in this village think that I am a witch*'. She started crying. I asked her—'why do people think that you are a witch?' She replied—'*I became a widow before my hair became white, so people believe that I am a witch*'.

'I am really scared to live here. Any day I may be beaten to death. No one in the village ready to help me'. Worrying about all these, she is not able to sleep peacefully. Her health has deteriorated and she shows all symptoms of severe depression.

Mrs. X. was considered as a 'witch' by upper caste families of the village. People do not want see them. They are not allowed to attend auspicious ceremonies. In all respects, they are socially and economically 'out-casted'. It is a pity that even educated persons in the village believe in this 'inhuman' practice of branding and treating old widows belonging to lower castes as witches.

Case Study 2

Mrs. C is a widow aged 76 years. She got married when she was 14. At the age of 34, she became a widow. She has one son and three daughters. When her husband died, her younger daughter was only 2 years old. She got 85 *Bighas* of land from her father at the time of marriage. Her husband was managing a petty shop in the village, and cultivating their land. She sold the land for meeting the treatment cost of her husband.

After his death, her condition became very pathetic. Somehow, she managed for few years. Her three daughters and son got married. After 2 years of her son's marriage, daughter-in-law started quarreling with her. She never gives her enough food. She deliberately wants to pick up quarrel with her mother-in-law. She slowly started humiliating her mother-in-law in public. Mrs. C was forced to work as a sweeper for 10 years. Her two daughters were staying in the nearby villages and one was in the same village. One day her son and daughter-in-law ordered her to leave their house. This was a shock to Mrs. C. After that, she started living with her daughter who lives in the same village.

She said—'*If you have money, then people will ask about you. But if you have nothing, then nobody will even come to see you*'. She continued—'*every parent hope that their sons will take care of them in old age. It is better to have a daughter rather than a son. If I do not have a daughter, then I will be begging somewhere now*'.

After the sudden death of her husband, she suffered a lot. She said about her son —'*For whom I was working so hard, threw me out when I really needed his help*'. She still expects love and care from her son. '*What I am expecting from my son may happen only in my next birth*'. Her son and daughter-in-law stay in the same village but they never come to see her even when she was sick. Mrs. C. had no option but to stay with her daughter, though it is not common in villages to live with son-in-law. '*When I was young, I never thought that one day I will be living in my son-in-law's house. What can I do? 'Every day I pray to God that I do not want to live long. How long one can suffer all these insults?*'

The frustration and helplessness in a widow's life is visible very much. All her hopes were shattered when her son and daughter-in-law started abusing her. Since then, she never had happiness and peace of mind. '*Well, we expect many good things in life, but ultimately we experience many bad things*'—at the fag end of her life, this widow has become philosophical.

References

Alam, Moneer. 2006. *Ageing in India: Socio-economic and health dimensions*. New Delhi: Academic Foundation.

American Psychological Association. 2004. *Elder abuse and neglect: In search of solutions*. www.apa.org/ageing/elderabuse.html.

Berkman, L., T.V. Sekher, B. Capistrant, and Y. Zheng. 2012. Social networks, family, and care giving among older adults in India. In *Aging in Asia: Findings from new and emerging data initiatives*, ed. J.P. Smith and M. Majmundar, Committee on Population, Division of Behavioral and Social Sciences and Education. Washington, DC: The National Academies Press.

Bhatt, Vighnesh N. 2006. Status of widows in Karnataka: Historical account and analysis of present scenario. In *Widowhood in modern India*, ed. P.K.B. Nayar. New Delhi: The Women Press.

Carolyn, Balkwell. 1981. Transition to widowhood: A review of the literature. *Economic and Political Weekly* 30: 117–127.

Census of India. 2011. *Social and cultural tables India*. New Delhi: Registrar General and Census Commissioner of India.

Chen, Martha Alter (ed.). 1998. *Widows in India: Social neglect and public action*. New Delhi: Sage Publications.

Government of India. 1999. *National policy on older persons*. New Delhi: Ministry of Social Justice and Empowerment.

Greg, Arling T. 1976. The elderly widows and her family, neighbours and friends. *Journal of Marriage and the Family* 38: 757–768.

Gupta, I., and D. Sankar. 2002. *Health of the elderly in India: A multivariate analysis*. Discussion paper 4. New Delhi: Institute of Economic Growth.

Gupta, S.K. 2017. *Elderly widows in India*. Jaipur: Rawat Publications.

Gupta, S.K., and T.V. Sekher. 2017. Vulnerability and coping mechanism of aged: A study of elderly widows in Jharkhand. In *Elderly care in India: Societal and state responses*, ed. S. Irudaya Rajan, and G. Balagopal, 169–192. Singapore: Springer.

Kumari, S., and T.V. Sekher. 2010. Elderly in Punjab: Economic dependency, health conditions and living arrangements. *Man and Development* 32: 77–88.

Kumari, S., and T.V. Sekher. 2012. Status of elderly widows in India. *BOLD* 22: 6–13.

Pandya, R., and S. Priti. 2006. Problems of elderly widows of Baroda-Gujarat. In *Older persons in India*, ed. Arvind K. Joshi. New Delhi: Serials Publications.

Rajan, S.Irudaya, and G. Balagopal (eds.). 2017. *Elderly care in India: Societal and state responses*. Singapore: Springer.

Sebastian, D. 2011. Abuse and neglect of the elderly in Indian families. *BOLD* 21: 5–12.

Sebastian, D., and T.V. Sekher. 2011. Extent and nature of elder abuse in Indian families: A study in Kerala. *Research and Development Journal* 17: 20–28.

Shankardass, M.K. 2010. Elder abuse and adult protection. In *Ageing and health in India*, ed. S. Johnson, and S.I. Rajan, 187–209. Jaipur: Rawat Publications.

Sivagami, A., and M. Thamilrasan. 2004. Empowering widows for development. In *Problems of widows in India*, ed. P.Adinarayana Reddy. New Delhi: Sarup & Sons Publications.

Troisi, J. 2000. *Tribal religion*. New Delhi: Manohar Publishers.

UNFPA. 1998. *The state of world population*. New York: United Nations Population Fund.

United Nations. 2001. *Widowhood*: invisible women, secluded or excluded, women 2000 and beyond, division for the advancement of women, united nations.

United Nations. 2002. *Gender dimensions of ageing: Women*. New York: Department of Economic and Social Affairs.

Yadava, K.N.S. 2011. Attitude and behaviour of family members and satisfaction of elderly: Some related issues. In *Ageing-some emerging issues: Profiles, trends and policy perspectives*, ed. K. N.S. Yadava, and Alok Kumar. New Delhi: Manak Publications.

Chapter 10
Are Elderly People Safe in Their Own Households? New Evidence from Seven States of India

Daliya Sebastian and T.V. Sekher

Background

With advances in the areas of health-care, nutrition and sanitation, the longevity of people has shown significant improvement around the world. Form the mid-twentieth century, the proportion of the aged population (60 years and above) has shown steady and significant increases in developed countries, and began to increase in developing countries based on the phase of demographic transition. Population projections indicate that by 2050, around 21% of the world's population will be elderly, and the proportion varies from 32% in developed countries to 19% in developing countries (United Nations 2013).

The dramatic and widespread nature of ongoing demographic shifts indicates that the challenges of ageing population that India will face are sure to occur on an enormous scale. India is currently experiencing an unexpected wave of demographic changes. Increasing longevity and falling fertility have resulted in a demographic dividend, but also in a dramatic increase of the elderly population. Along with an increase in the proportion of older people, the old-age dependency ratio (population age 65+/population age 25–64) will also increase at rapid a scale and, according to United Nations estimates, the old-age dependency ratio will increase from the present 8.6 to 20.5% by 2050.

The elderly population in India has increased steadily from 12 million in 1901 to more than 100 million in 2011 (Fig. 10.1). Population projections indicate that India's elderly population will double in size between 2001 and 2026 (MoS&PI 2011). One of the significant characteristics of population ageing will be the wide

D. Sebastian (✉)
International Centre for Research on Women (ICRW), New Delhi, India
e-mail: daliasebastian@gmail.com

T.V. Sekher
International Institute of Population Sciences (IIPS), Mumbai, India
e-mail: tvsekher@gmail.com

Mala Kapur Shankardass and S. Irudaya Rajan (eds.), *Abuse and Neglect of the Elderly in India*, https://doi.org/10.1007/978-981-10-6116-5_10

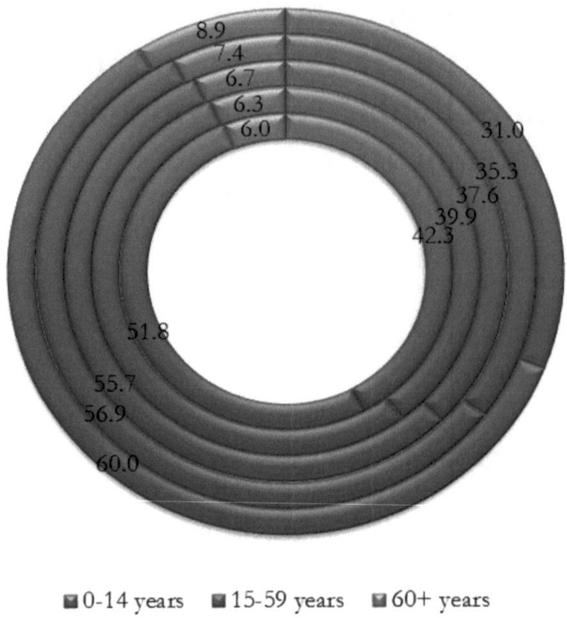

Fig. 10.1 Age distribution of population by broad age groups in India, 1971–2011

inter-state variation based on the demographic scenario. As per the projections, by 2026, the North Indian population will be younger than that in the South. The median age of Kerala will be 35 while that in Uttar Pradesh will be less than 30 years (MoS&PI 2011).

While the numbers and proportion of elderly people have gone up, the vulnerability of the ageing population is also on the rise. Urbanization and migration have accelerated issues of living arrangements, economic situation, and abuse and neglect of elderly people in both developed and developing countries. The available literature from developed countries indicates that even with a strong social support mechanism for the elderly, a large number among them are victims of abuse and neglect from family members, informal care-givers, friends or formal care-givers. It also reveals that over the decades, the proportion of elderly people who are victims of abuse and neglect is also increasing (American Psychological Association 2004). The extent of elder abuse ranges considerably across the populations. According to a study conducted by King's College London, it is estimated that the prevalence of abuse among the elderly residing in private households was around 2.6% in 2008. Findings from Argentina reported a higher proportion of elder abuse, with nearly 45% of the elderly from urban areas reporting that they were abused in some way (United Nations 2002). A systematic review of existing studies of elder abuse and neglect across the world observed that the prevalence of elder abuse ranges from 3.2 to 27% in the general population. A study by the Korea Institute for Health and Social Affairs found that nearly 8% had experienced abuse from their children or other family members in Korea (Cooper et al. 2008).

While large variations are observed in the extent of elder abuse in different studies, the type of abuse reported also varies widely. However, the majority of the studies reported neglect or verbal abuse as the most commonly cited forms (Douglass et al. 1980; Comijs et al. 1998; National Elder Abuse Study 1998; UN 2002). However, there is also evidence that physical abuse is the most prevalent form (Pillemer and Finkelhor 1987; Arai 2006). The national elder mistreatment study conducted in 2009 reported 31% of physical abuse among elderly people in the USA compared to 8% experiencing emotional mistreatment (Acierno et al. 2009).

Regarding the risk factors of elder abuse and neglect, there is evidence that women are at higher risk of being abused or neglected than men (APS 2004; UK Study 2004; Eriksson 2001). On the other hand, there are studies which indicate that gender of the elderly is not a risk factor (NCEA 1998; Dimah and Dimah 2002). Most of the studies reported that elderly in advanced age especially those who are aged 80 and above, and elderly people who are economically dependent on care-givers for their basic needs are more vulnerable to abuse and neglect (Pillemer and Finkelhor 1987; NCEA 1998). A study which examined the explanations for the occurrence of elder abuse found that, among multiple explanations, the predominant factors are economic dependency, modernization, and social integration (Litwin and Zoabi 2004).

Abuse and neglect of the elderly were not considered problems in Indian society till recently, with the wide existence of the joint-family system, where elderly parents were treated with great respect and played a major role in family decisions. The elderly were largely taken care of by their immediate family, mostly by their sons in India. However, in recent times, with economic and social transition, we are experiencing a shift in the family structure as well and the joint-family system is crumbling not only in urban areas but also in rural India. The disintegration of joint-families coupled with fewer children to depend during old age and high migration of youngsters in search of better employment has left an increased number of their parents alone and vulnerable to abuse and neglect. Unlike developed countries, the social security system and government support for the elderly who are in need is not yet available in India. As per the Global Age Watch Index (2013), India ranked 73 out of 91 countries, based on four domains of income security, health status, employment and education and enabling environment (HelpAge International 2013).

The problems of the aged, especially abuse and neglect, received very little attention in India. The researches on ageing were mostly confined to seniors' health problems, economic conditions and living arrangements. Most of the available studies on elder abuse and neglect were largely limited to developed countries. However, during the last decade there has been an increased interest in studying elder abuse and neglect in developing countries. Still, little is known about its magnitude and nature in the family context, and the majority of the studies depend on qualitative information at micro level and anecdotal evidence (Jain 2008; Madhurima 2008). Given this context, there is an increasing need for nationally representative data and evidence on elder abuse.

The reported prevalence of elder abuse in India varies in different studies. According to a study conducted among 400 elderly people aged 65 years and above in Chennai, 14% were abused (Chokkanathan and Lee 2006) while the prevalence of abuse reported among the elderly from Delhi was much higher at 31% (Khan and Handa 2006). Another study reported nearly 49% of the elderly respondents from rural areas of Kerala experienced some form of abuse and neglect within households (Sebastian and Sekher 2011). Considering the social customs prevailing in the country, the reported cases will be showing only the tip of the iceberg. Based on the information provided by 100 elderly women from Pune, nearly a quarter faced abuse and neglect in their homes, and the majority reported more than one form of abuse (Bagga and Sakurkar 2011). A content analysis of newspaper stories concerning various facets of elder abuse found that 10 out of 16 newspapers reported evidence of elder abuse and neglect across Indian cities (Kumar and Bhargava 2014). While exploring the most common forms of abuse faced by the elderly in India, verbal abuse and neglect were the most cited forms in the majority of the studies (Srinivas and Vijayalakshmi 2001; Chokkanathan and Lee 2006; Bagga and Sakurkar 2011; Sebastian and Sekher 2011). A study conducted by HelpAge India in 2013 also reported that the most cited form of abuse experienced by elderly people was disrespect followed by verbal abuse and neglect.

There are various factors which determine the extent and nature of abuse towards elderly people, considering the socio-economic conditions prevailing in the society. In India, females were considered as more disadvantaged in all spheres of life and the extent of vulnerability increases when they become old. Elderly women are more susceptible to abuse and violence when they are widowed. Review of existing studies indicates that sex and marital status of elderly people plays a significant role among the set of factors which accelerate the elder abuse and neglect. The existing studies provided clear evidence that elderly females, especially widows, were more vulnerable to abuse and neglect (Dejong 2004; Chokkanathan and Lee 2006; Madhurima 2008; Sebastian and Sekher 2011). Economic dependency and living arrangements also emerged as key factors of elder abuse and neglect.

HelpAge India's report on Elder Abuse ('The Indian Youth Speaks Out'), based on a study among youths from 10 cities of India in 2015 reveals that 73% admit that the abuse and neglect of elderly people does exist in Indian society. It was also reported that although they were aware of the issues of elder abuse, the majority of them do not care or do not want to interfere/are not willing to take action against such incidences. Alarmingly, 35% of the youths reported that they had noticed elder abuse happening among their own relatives or in their own households (HelpAge India 2015).

HelpAge India carried out a survey in 8 cities, covering 833 respondents, from the oldest old (80+ years in age) in Delhi and the National Capital Region, Mumbai, Hyderabad, Chennai, Patna, Kolkata, Bhopal and Ahmedabad in 2014. According to this study, one-fifth of them reported having experienced some kind of abuse. The types of abuses reported were mostly verbal abuse, followed by neglect, disrespect, emotional and physical abuse (HelpAge India 2014).

The data from the Building Knowledge-Base on Population Ageing in India (BKPAI) Survey (2011) found that older females and the financially dependent older adults are more likely to face abuses of different kinds. The existing evidence also revealed that after inheriting property, the children behaved differently and started neglecting the needs of parents (Sebastian 2013). Skirbekk and James have explored the role of education and found that the level of schooling among elderly people is strongly negatively related to abuse against them. Even after controlling for wealth and other relevant variables, education is the factor that most consistently lowers elderly abuse (Skirbekk and James 2014).

In most of the cases, the main perpetrators are the immediate family members and that aggravates the helplessness of the elderly as they were reluctant to report the abuse and neglect. Sons and daughters-in-law are the main perpetrators of elder abuse in most of the cases, since they are the primary care-givers of elderly people (Sebastian and Sekher 2011).

Although various studies indicate that the abuse and neglect of elderly people do exist in our society, a need is felt to study their prevalence. The main objective of this chapter is to understand the magnitude of elder abuse in India and the factors influencing it, based on the data of seven Indian states.

Methodology

The present chapter uses the data from the Building a Knowledge-Base on Population Ageing in India (BKPAI) survey which was conducted in seven states of India—Himachal Pradesh, Maharashtra, Kerala, Punjab, Odisha, Tamil Nadu, and West Bengal—in 2011 by the United Nations Population Fund (UNFPA) in collaboration with the Institute of Social and Economic Change (ISEC), Bangalore, the Institute of Economic Growth, Delhi, and the Tata Institute of Social Sciences (TISS), Mumbai. The states were selected considering the share of the elderly population and regional representation. A total of 9852 elderly people were interviewed covering 8960 households (see Table 10.1).

The information was collected using a household schedule which covered the socio-economic and demographic information of the household and the type of amenities to measure the living standards of elderly people. In addition to the household schedule, the individual schedule for the elderly was canvassed to capture the demographic profile, work status, living arrangements, economic conditions, health and health-seeking behaviour, family situation including elderly abuse and neglect, and awareness about social security schemes etc.

By using the individual data of elderly people from the BKPAI survey, this chapter looks at the extent of abuse and neglect experienced by seniors within their families or outside. It also tries to examine the various contributing factors which led to the abuse and neglect of elderly in Indian settings. Univariate, bivariate and multivariate analyses were conducted to understand the magnitude and nature of abuse reported.

Table 10.1 Demographic profile of study states

Indicators	India	HP	Kerala	Maharashtra	Odisha	Punjab	TN	WB
Population in millions (2011)	1210.9	6.9	33.4	112.4	42.0	27.7	72.1	91.3
Exponential growth rate of population (%)	1.64	1.21	0.48	1.49	1.32	1.30	1.46	1.31
Percentage of elderly (60+)	8.9	10.2	12.6	9.9	9.5	10.3	10.4	8.5
Sex ratio (F/M * 1000)	943	972	1084	929	979	895	996	950
Sex ratio among elderly (F/M * 1000)	1033	1062	1226	1114	998	985	1051	1010
TFR, lifetime births per woman (2013)	2.3	1.7	1.8	1.8	2.1	1.7	1.7	1.6
Life expectancy at birth in years (2009–2013)	67.5	71.0	74.8	71.3	64.8	69.9	70.2	69.9

Sources Registrar General of India: Census, 2011, SRS reports 2013, SRS based abridged life table (2009–2013)
HP Himachal Pradesh, *TN* Tamil Nadu, *WB* West Bengal

Profile of the Respondents

Out of 9852 elderly people who participated in the study, 52.6% were females and 52.2% from rural areas (see Table 10.2). Looking at the age distribution, nearly 63% of respondents were in the age category of young old (60–69 years), 26% were old old (70–79 years) and 10% were oldest old (80 years and above). The majority of the respondents followed the Hindu religion and 24% belonged to the Scheduled Caste/Scheduled Tribe (SC/ST) category. A little more than half of the respondents reported they had ever attended school. Nearly 60% of the elderly respondents were currently married and 38% were widowed. Some 36% of the respondents reported they were engaged in income-generating activity during the last year prior to the survey. The major reason cited for working during old age was economic need (67%) and 30% reported that they worked out of choice. More than half of the elderly who were engaged in economic activity experienced physical or mental stress due to the work.

Table 10.2 Background characteristics of elderly respondents

Background variables	Percentage (numbers)
Sex	
Male	47.4 (4672)
Female	52.6 (5180)
Age group	
Young old (60–69 years)	63.3 (6239)
Old old (70–79 years)	26.4 (2601)
Oldest old (80 years and above)	10.3 (1012)
Place of residence	
Rural	52.2 (5138)
Urban	47.8 (4714)
Religion	
Hindu	79.0 (7781)
Others	21.0 (2071)
Caste	
SC/ST	24.8 (2383)
OBC	34.0 (3353)
Others	40.3 (3866)
Ever attended school	
Yes	54.0 (5324)
No	46.0 (4528)
Marital status	
Married	59.3 (5847)
Widowed	38.2 (3768)
Others	2.5 (237)
Worked during last year	
Yes	36.1 (2265)
No	63.9 (4001)
Reason for working	
By choice	29.1 (660)
Economic necessity	66.1 (1498)
Other compulsions	4.7 (107)

Living Arrangements of Elderly People

Elderly people's living arrangements largely depend upon the existing cultural norms and the support mechanism, especially the inter-generational support system available to them. In India, where seniors have limited or no resources to sustain themselves and lack a social support mechanism, they largely depend on their heirs for support during old age. Living independently, especially living alone, could be a disadvantage or even a result of neglect.

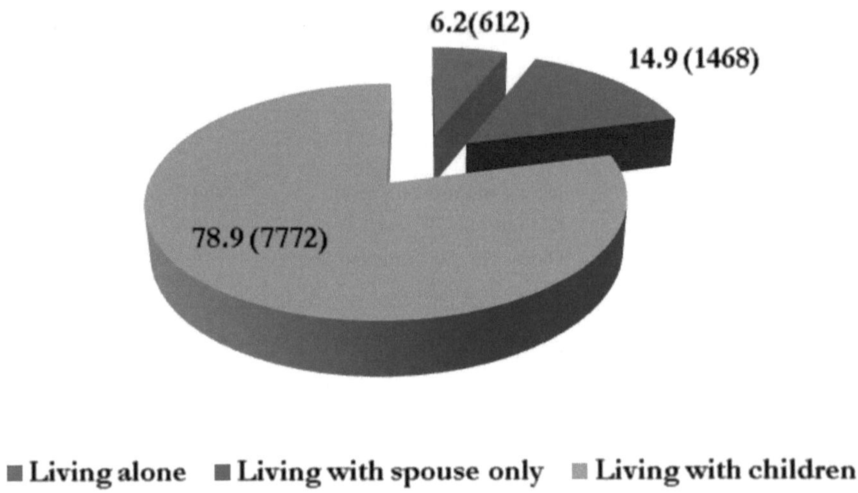

Fig. 10.2 Living arrangements of elderly people

Nearly 79% of respondents were living with their children and family, 15% were living only with a spouse and with or without a servant, and 6% were living alone (Fig. 10.2). Considerable inter-state variations were observed in the living arrangements of elderly people, with nearly 44% of those from Tamil Nadu reporting either that they were living alone or only with a spouse (15.8 and 27.7% respectively) compared to around 15% in Kerala and Punjab.

Of those who were living alone or only with a spouse, nearly half reported children living away as the reason for their current residence pattern, while 16% reported family conflict as the reason. Nearly 14% do not have any children and are residing alone or with spouse only. Overall, 40% of the respondents stated that they were comfortable with present living arrangements, and 48% found them satisfactory, while about 12% were uncomfortable. A higher proportion of elderly people living alone reported being uncomfortable (27%) than those who were living with their children and family (10%).

Elderly Involvement in Decision Making Within the Family

The ongoing transformation of Indian families from joint-families to nuclear ones has also greatly affected the role of elderly people's decision-making power within the family. With less or no involvement in the decision making of family-related matters, seniors felt neglected and deprived. The analysis indicates that even though nearly two-thirds of the elderly reported there is no change in their role as decisionmakers after they become 60 years old, around 26% felt that their role as a decisionmaker declined after they became aged. A higher proportion of elderly people from Tamil Nadu (52%), West Bengal (44%) and Odisha (40%) reported decline in their decision-making capacity than those from other states (where the figure varied between 10 and 13%). As expected, a higher proportion of females from all states reported a decline in their decision-making capacity when they became aged.

Abuse and Neglect of Elderly People

The information about the abuse and neglect experienced by the elderly was collected from 9779 respondents (in 73 cases, the respondents were not able to respond to the questions due to health issues). The information on abuse and disrespect was collected from seniors ensuring privacy and non-presence of other family members to gather accurate information.

Conceptualization of Different Forms of Elder Abuse

Physical Abuse

This includes beating, kicking, slapping, stabbing, shooting, pushing, biting, pinching, strangling, and so on.

Verbal Abuse

This is a form of abusive behaviour involving the use of language. It is a form of vulgarity that can occur with or without the use of expletives. It can be either through oral communication, which is the most common form of verbal abuse, or abusive words in written form.

Economic Abuse

This involves the illegal or improper use of a senior's funds, property, or assets. Examples include, but are not limited to, cashing an elderly person's cheques without authorization or permission, forging an older person's signature in documents, and misusing or stealing an older person's money or possessions, coercing or deceiving an older person into signing any document (e.g., contracts or will), and so on.

Showing Disrespect

It is expected for people to show proper respect to elders, particularly in Indian culture. Any act or behaviour showing lack of respect to a senior will come under this category.

Neglect

The failure for a care-giver to meet the needs of a dependent elderly person (which may be intentional such as withholding of food, medications, failure to clean or bathe, and so on) comes under this category.

Nearly 10% of the elderly interviewed reported they had experienced abuse or disrespect. Considerable inter-state variations were observed among the elderly who reported abuse, ranging from with nearly 31% in Maharashtra and only 2% in Tamil Nadu (Fig. 10.3). Among the elderly who experienced abuse and disrespect, nearly 35% experienced multiple abuse. A higher proportion of elderly people from Odisha (60%) and Punjab (55.6%) reported multiple abuses.

When elderly people reported having experienced abuse or disrespect, further probing was done to understand the nature of that abuse or disrespect and where it had taken place (Fig. 10.4). Data were collected on incidents of physical abuse, verbal abuse, economic abuse, neglect and disrespect. Verbal abuse was the most cited form of abuse experienced by seniors (89.5%), followed by disrespect

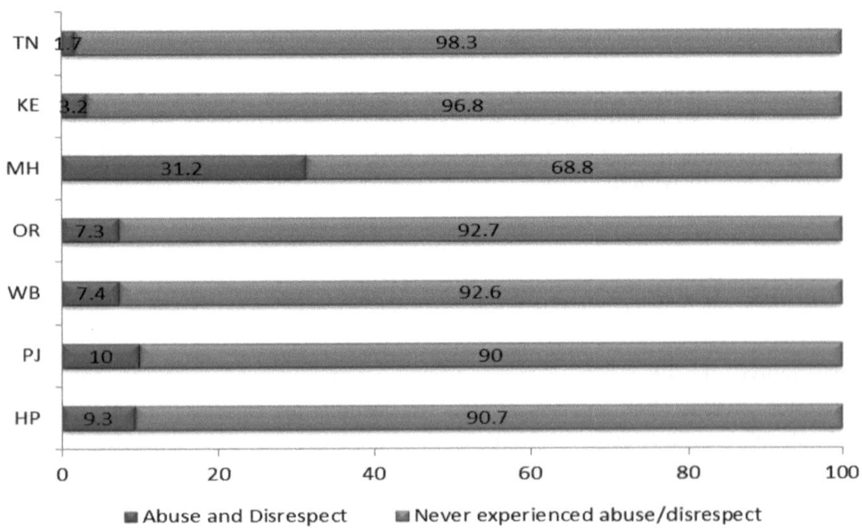

Fig. 10.3 Proportion of elderly people reporting abuse and neglect by states

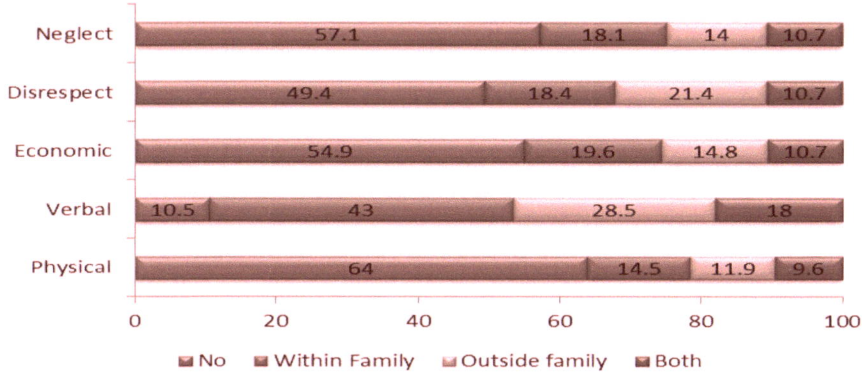

Fig. 10.4 Percentage of elderly people who reported abuse by type and place of abuse

(50.6%), economic abuse (45.1%), neglect (42.9%) and physical abuse (36%). It was also noted that the majority of them experienced the abuse and neglect within family rather than outside the family. Of those who reported that they had been abused or mistreated, nearly 50% said it had occurred during the month prior to the survey. Of those who reported abuse and disrespect during the one month prior to the survey, 87% reported that they had experienced verbal abuse, followed by disrespect (35%) and neglect and economic abuse (31%). Nearly 29% of the respondents reported that they were victims of physical abuse during the month prior to the survey. Some 22% of the seniors who had experienced abuse and neglect during the month prior to the survey reported incidences of multiple abuse.

Abuse and neglect of elderly people have consequences for victims' physical and mental health and lives as a whole. Earlier studies have indicated that depression among elderly people who experienced abuse as much more severe than depression among those who had never experienced abuse. Among the respondents who had experienced abuse and disrespect during the month prior to the survey, nearly 22% reported that they were suffering from health problems.

Contributing Factors for Abuse and Neglect

An attempt was made to understand the contributing factors which led to abuse and neglect of elderly people within the family as well as outside homes (Table 10.3). With increase in age, the extent of abuse and neglect towards elderly people is also increasing. Around 9% of young old respondents (60–69 years) reported abuse and neglect, and the proportion increased to 14% among the oldest old (80 years and above). A slightly higher proportion of females (10.6%) reported abuse and disrespect than their male counterparts (9.5%). It was also noted that the lower proportion of elderly people who were currently married reported abuse and neglect than those who had lost their spouses. The living arrangement of elderly people also

Table 10.3 Percentage of elderly people reporting abuse and disrespect, by background characteristics

Background characteristics	Percentage (numbers)
Faced abuse/disrespect	10.1 (987)
State	
Himachal Pradesh	9.3 (136)
Punjab	10.0 (133)
West Bengal	7.4 (94)
Orissa	7.3 (108)
Maharashtra	31.2 (448)
Kerala	3.2 (43)
Tamil Nadu	1.7 (25)
Sex	
Male	9.5 (440)
Female	10.6 (547)
Age group	
Young old	9.4 (581)
Old old	10.1 (262)
Oldest old	14.4 (144)
Marital status	
Currently married	9.1 (529)
Widowed	11.3 (423)
Other	15.0 (35)
Living arrangements	
Alone	16.6 (101)
With spouse only	10.6 (155)
With children	9.5 (731)

showed a significant effect on the abuse and neglect, with those who were living alone reporting higher incidences of abuse and neglect.

The logistic regression analysis indicates that the elderly who are living alone and those who belong to rural areas were more likely to be victims of abuse and neglect during their advanced ages. The elderly having 8 or more years of schooling were 50% less likely to be abused compared to the non-literate seniors. Economic dependency of elderly people towards the care-givers and the number of living sons are the other significant factors contributing to the abuse and neglect of elderly people in the family and in the community (Table 10.4).

Perpetrators of Abuse

Earlier studies have indicated that children, especially sons and daughters-in-law, were the main perpetrators of abuse of seniors. Most of the time, there were multiple perpetrators involved in the abuse and neglect of elderly. This study also

Table 10.4 Results of logistic regression on abuse and disrespect of seniors

Characteristics	Category	Odds ratio (95% CI)	*p* value
Sex of the respondent	Males (ref)		
	Females	0.913 (0.711–1.172)	0.474
Age group of the respondent	60–69 years (ref)		
	70 years and above	0.901 (0.738–1.100)	0.306
Marital status	Married (ref)		
	Other	1.150 (0.903–1.465)	0.257
Current living arrangement	Living alone/with spouse only (ref)		
	Living with children	0.487* (0.394–0.600)	0.000
Place of residence	Rural (ref)		
	Urban	0.748** (0.616–0.910)	0.004
Number of years of schooling	Non-literate (ref)		
	1–4 years of schooling	1.061 (0.810–1.388)	0.669
	5–7 years of schooling	0.941 (0.714–1.241)	0.668
	8 and more years of schooling	0.483* (0.365–0.638)	0.000
Economic dependency on basic needs	Fully dependent (ref)		
	Partially dependent	0.613* (0.474–0.791)	0.000
	Not dependent	0.548* (0.417–0.720)	0.000
Number of living sons	No sons (ref)		
	One son	1.296 (0.932–1.803)	0.123
	More than one son	1.721** (1.258–2.355)	0.001
Current work status	Not working (ref)		
	Working	1.429** (1.151–1.774)	0.001

shows the same trend, with nearly 45% of seniors who experienced abuse and neglect during the month prior to the survey reporting that there were multiple perpetrators. The main perpetrators were neighbours (46%), sons (40.7) and relatives (32%). One in ten elderly people reported that they had faced abuse from grandchildren. The majority of them reported that they had faced abuse and disrespect within the household, especially verbal abuse and neglect.

Concluding Observations

Elder abuse and neglect are increasingly acknowledged as social problems internationally, and India is no exception. The responsibility for caring for the elderly in India is traditionally borne by the immediate family. The society's transition, with more and more people opting to go out of their local habitations to work and lead a nuclear family has led to more and more elderly people living alone. Inter-state and inter-country migrations are leading to both physiological and psychological vulnerability. Though a majority of elderly people interviewed have not experienced any form of abuse after they turned 60 years of age, one in ten reported facing abuse within the household or outside. The mistreatment of elderly people is multi-dimensional and multi-layered, emerging from the differences in gender, economic position, and physical condition (Berkman et al. 2012). Economic dependency and low education seem to be the important factors making the aged more susceptible to abuse. The elderly, who reside in urban areas, are less likely to be abused compared to those in rural areas. Those seniors living with children are less likely to be abused than those who are living alone or with a spouse. Higher levels of education and financial security during old age seem to be the factors that make the elderly less vulnerable to abuse. This study indicates that the main perpetrators of abuse were from outside the family for elderly men and within the family for elderly women. Maintaining the traditional family values of love, caring, and respect for seniors in recognition of their contribution and experience would be the most important areas upon to be focused in future, which could bring in the much needed cohesion in the family, making older people feel more valued and less susceptible to abuse and violence. Though the government has come out with the Maintenance and Welfare Parents and Senior Citizens Act in 2007, very few elderly people are aware of it and the Act is not yet fully operational in most states. Awareness campaigns are needed to educate older persons about their rights and entitlements.

The elderly are in need of a wide range of supportive, preventive and protective services. It is the primary responsibility of the family and children to provide love, support and care to its older members. However, in many instances, seniors are vulnerable to mistreatment by their family members and care-takers. To some extent, the evidence emerging out of this study has questioned the general perception in India that the families are the safest place for the aged to live. A combination of governmental, non-governmental and community efforts are required to ensure that seniors spend their remaining years in a peaceful and dignified manner.

Annex: Case Studies

Case Study 1

Mr. Y is an 83-year-old widower, living with his youngest son and family. He belongs to an aristocratic Christian family. He has 10 children—seven sons and three daughters—and all of them are well settled in different parts of India and abroad. His wife passed away a few years back. He had around 50 acres of land and divided this between all his children and has kept only four acres of land and house in his own name. The youngest son is living with him and takes care of him properly. He used to enquire daily about his food requirements and instruct his wife to give milk to Daddy. The other sons were not so happy with Mr. Y because he had given a major share of the land to the youngest son. He justified his action by saying 'I have given 15 acres of land to him with the expectation that he will take care of me'. Mr. Y is an asthma patient and suffers from diabetes and high blood pressure. Whenever he asked his son to take him to a hospital, he used to oblige. Recently, his son told him to write a will stating that, after his death, the property (4 acres of land and house) would be for him along. Mr. Y called the Registrar to his home and signed the required documents.

Though he is fond of his youngest son, sometimes he feels that he is not getting enough care and attention at home. 'Whenever I ask for any specific need, he is very reluctant to help. Maybe he is too busy with so much of work'. But Mr. Y narrated some incidents when he had felt very upset. 'One day, when my daughter-in-law asked me to have lunch, I was not feeling hungry, so I told her I will come after one hour'. After one hour, when I asked for lunch, she replied like this: 'I am not running a restaurant here to supply food at any time'. When I told my son about this, he supported his wife by saying, 'She has a lot of work, so you try to take food on time. Even if you do not take food for one day, nothing will happen to you'. When Mr. Y wanted his son to take him to a hospital, the reaction was, 'Today I don't have time. During our last visit, the doctor told me that you don't have any health problem. It is only a psychological problem'. When my elder son came to see me after many years, I informed all these things. He just replied, 'It is your mistake. Who told you to execute such a will and hide everything from all of us? Now we can't do anything'. Last month, he was admitted to hospital due to high blood pressure and asthma. His youngest son told his brothers, '"I cannot stay in hospital with him. So any one of you please stays with him or arranges some servant to take care of him at home". But my other sons are also not interested in taking care of him'. His second daughter told him that he can stay with her family. 'It is a shame to go and stay with daughter after having seven sons.' The old man concluded, 'It was my mistake that I gave more love and property to my youngest son. Now I have to suffer silently and nobody else will take care of me'. He now realizes that his judgement was wrong and he blames himself for the situation. At least, he expected a sympathetic attitude and kind words from children. Mr. Y is not fortunate enough to have that in the last years of his life.

Case Study 2

Mrs. A is a widow from a very poor family, living with her son and daughter-in-law. She has one son and one daughter. She worked as a part-time servant in a house and has stopped working for the last five years due to her health problems. Her son is a heart patient. He is not able to do any work. Her daughter-in-law is working as an agricultural labourer and her earning is the only source of income for the family. While describing the behaviour of her daughter-in-law, she said 'My daughter-in-law is not talking with me for many years, even though we are living in the same house. She never called me "Amma" (mother) during last seven years and she uses some offensive slang word instead of "Amma" while mentioning about me to others. She leaves the food in my room and never tells me whether the food is ready or not. Even if I am sleeping she leaves the food in the room and goes away'. Since she is the only breadwinner of the house, my son never questions her on anything. Moreover, he is in a depressed condition due to his illness and economic problems. She is also tense about the attitude of her own daughter who is married to a family in the same village. In her words, 'If my own daughter is not interested to see the old mother, what can we expect from daughter-in-law?' Mrs. A spent all her little earnings on her daughter's marriage. Nowadays, the daughter has stopped coming to meet her and will not help in any way even though she is in a better-off condition. After much probing, the old widow explained how she was treated and neglected in her family. 'One day I fell down in the courtyard and fractured my leg. Somehow, I managed to come up to the house before fainting. When I got up, I saw no-one in the house. So I screamed and one of my neighbours informed my daughter about my injury. When she was informed about my fracture, her response was, 'I have enough responsibilities in my home and I don't have time to take care of her. Moreover, it is not my duty. It is the duty of my brother as he got the house from mother'. My neighbours took me to the hospital and provided me with food for one month. My son and daughter-in-law came to see me after a month. My daughter-in-law removed the cot and mattress from my room. Now two wooden planks are kept on the bricks as a bed. 'Nowadays, my daughter-in-law does not give me enough food to eat. If I protest, she will shout at me.' Mrs. A is totally depressed and worried: 'I don't have any economic assets; I spent all my earnings for on daughter's marriage and gave the land and house to my son. So I have to depend on them for my daily requirements. I am afraid to commit suicide. I do not have courage. I am praying to God to take me away from this world at the earliest. Why should I suffer and live like a beggar?'

It was very difficult to interview Mrs. A on the first visit as her daughter-in-law was observing everything. During the next visit, fortunately her daughter-in-law was away and I could talk to the old lady in a relaxed manner. Though her son was at home, he never asked anything. When I was saying goodbye after a long chat, the old lady told me, 'I hope you will take care of your mother. Please do not behave like my daughter-in-law'. She also reminded me, 'What I discussed with you is between you and me. Never mention this to my daughter-in-law. If she happens to know, she will throw me out and next time you will see me lying on the street'.

Note: The above two case studies were part of a study on elder abuse conducted in rural Kerala (Sebastian 2013).

References

Acierno, R., M.H. Tejada, W. Muzzy, and K. Steve. 2009. National elder mistreatment study. https://www.ncjrs.gov/pdffiles1/nij/grants/226456. Accessed 24 January 2016.

American Psychological Association. 2004. *Elder abuse and neglect: In search of solutions*. www.apa.org/ageing/elderabuse.html. Accessed 24 January 2016.

Arai, M. 2006. Elder abuse in Japan. *Educational Gerontology* 32: 13–22.

Bakshi, S., and P. Pathak. 2016. Aging and the socioeconomic life of older adults in India: An empirical exposition. *SAGE Open*. doi:10.1177/2158244015624130.

Berkman, L., T.V. Sekher, B. Capistrant, and Y. Zheng. 2012. Social networks, family, and care giving among older adults in India. In *Ageing in Asia: Findings from new and emerging data initiatives*, ed. J.P. Smith, and M. Majmundar. Washington, DC: National Academies Press.

Chokkanathan, Srinivasan, and Alex E.Y. Lee. 2006. Elder mistreatment in urban India: A community based study. *Journal of Elder Abuse and Neglect* 17: 45–61.

Comijs, H., J. Smith, A. Pott, L. Bouter, and C. Jonker. 1998. Risk indicators of elder mistreatment in the community. *Journal of Elder Abuse and Neglect* 9: 67–76.

Cooper, C., A. Selwood, and G. Livingston. 2008. The prevalence of elder abuse and neglect—A systematic review. *Age and Ageing* 37: 151–160.

Dimah, Agber, and Keren Patricia Dimah. 2002. Gender differences among abused older African Americans and African–American abusers in an elder abuse provider agency. *Journal of Black Studies* 3: 557–573.

Douglass, R., T. Hickey, and C.A. Noel. 1980. *Study of maltreatment of the elderly and other vulnerable adults*. Ann Arbor, MI: Institute of Gerontology, University of Michigan.

Eriksson, H. 2001. *Violence against elderly women and men: A prevalence study*. Umea: Umea University Sweden.

Help Age India. 2015. Elder abuse: The Indian youth speaks out. Research report on National Survey of Youth Perspective on Elder Abuse, 2015. https://www.helpageindia.org/pdf/ElderAbuse-The%20Indian-Youth-Speaks-Out.pdf. Accessed 24 January 2016.

Help Age India. 2014. State of elderly in India 2014. https://www.helpageindia.org/images/pdf/state-elderly-india-2014.pdf.

Help Age International. 2013. Global Agewatch Index 2013. *Insight Report*. http://www.helpage.org/download/541300b365b65.

Jain, U.C. 2008. Elder abuse: Outcome of changing family dynamics. *Indian Journal of Gerontology* 22: 447–455.

Khan, A.M., and Smita Handa. 2006. Physical abuse of elderly in Indian context. *Indian Journal of Gerontology* 20: 235–249.

Korean Institute for Health and Social Affairs. 2001. The study of the definition establishment and the actual condition of family violence in Korea. http://www.kihasa.re.kr/data/rm98-04. Accessed 2 April 2010.

Kumar, Y., and A. Bhargava. 2014. Elder abuse in Indian families: Problems and preventive actions. *International Journal of Scientific and Research Publications*. http://www.ijsrp.org/research-paper-1014/ijsrp-p3498.pdf. Accessed 21 January 2016.

Madhurima, M. 2008. Elderly widows as victims of physical abuse: A qualitative study in the state of Punjab. *Indian Journal of Gerontology* 22: 501–514.

Mahajan, A., and M. Madhurima. 1995. *Family violence and abuse in India*. New Delhi: Deep and Deep Publication.

Ministry of Statistics & Programme Implementation. 2011. *Situation analysis of elderly in India*. New Delhi: Government of India.

National Centre on Elder Abuse. 1998. What is elder abuse: What are the major types of elder abuse? http://www.interinc.com/NCEA/Elder_Abuse/main.html. Accessed 24 January 2016.

Pillemer, K., and D. Finkelhor. 1987. Domestic violence against the elderly. Paper presented at the third conference on family violence, Durham, May 28–June 30.

Registrar General of India. 2013. *Sample registration system statistical report 2013*. New Delhi: Ministry of Home Affairs, Government of India.
Registrar General of India. 2014. *SRS based abridged life tables 2007–2013*. New Delhi: Ministry of Home Affairs, Government of India. http://www.censusindia.gov.in/vital_statistics/Appendix_SRS_Based_Life_Table.html. Accessed January 2016.
Sebastian, D. 2011. Abuse and neglect of elderly in Indian families. *BOLD* 21: 5–12.
Sebastian, D. 2013. *Ageing and elder abuse in Kerala*. New Delhi: Rawat Publications.
Sebastian, D., and T.V. Sekher. 2011. Extent and nature of elder abuse in families: A study in Kerala. *Helpage India-Research and Development Journal* 17: 20–28.
Skirbekk, V., and K.S. James. 2014. Abuse against elderly in India—The role of education. *BMC Public Health* 14: 336–342.
Srinivas, S., and B. Vijayalakshmi. 2001. Abuse and neglect of elderly in families. *Indian Journal of Social Work* 62: 465–478.
UK Study of Abuse and Neglect of Older People. 2007. "*Research findings*". Kings College London and National Centre for Social Research. www.natcen.ac.uk. Accessed 24 January 2016.
United Nations. 2002. *Elder abuse widespread and unprotected.* United Nations Department of Public Information. http://globalag.igc.org/waa2/articles/elderabuse.htm. Accessed 24 January 2016.
UNFPA. 2012. *Report on the status of elderly in selected states of India—2011. Building a knowledge base on population ageing in India.* New Delhi: UNFP.
United Nations. 2013. *World population ageing—2013*. New York: Department of Economic and Social Affairs, Population Division.

Chapter 11
Contextualizing Elder Abuse and Neglect in Institutional and Home Settings: Case Studies from India

Selim Jahangir, P.N.N. Nikhil, Ajay Bailey and Anindita Datta

Introduction

Elder abuse or mistreatment refers to 'intentional actions that cause harm or create serious risk of harm (whether or not harm is intended) to a vulnerable elder by a caregiver or other person who stands in a trust relationship to the elder or [...] failure by a caregiver to satisfy the elder's basic needs or to protect the elder from harm' (National Research Council 2003: 1). The World Health Organization (2008) used the term 'elder abuse', adopting the definition developed in 1995 by Action on Elder Abuse in the United Kingdom, and has defined elder abuse as 'a single, or repeated act, or lack of appropriate action, occurring within any relationship where there is an expectation of trust which causes harm or distress to an older person' (WHO 2008: 6). Elder abuse and neglect are increasingly recognized as emerging global social care and public health issues (Lachs and Pillemer 2004) and India is no exception (Sachan and Kaur 2014). Traditionally, the older adults' care responsibility was borne by the immediate family members with great respect and filial obligations in India. But due to rapid urbanization and modernization, the

S. Jahangir · A. Datta
Department of Geography, University of Delhi, New Delhi, India
e-mail: selimjahangir007@gmail.com

A. Datta
e-mail: anindita.dse@gmail.com

P.N.N. Nikhil
Population Research Centre, University of Groningen and Institute for Social and Economic Change, Bangalore, India
e-mail: nikhil@isec.ac.in

A. Bailey (✉)
Population Research Centre, Faculty of Spatial Sciences, University of Groningen, Groningen, The Netherlands
e-mail: a.bailey@rug.nl

Mala Kapur Shankardass and S. Irudaya Rajan (eds.), *Abuse and Neglect of the Elderly in India*, https://doi.org/10.1007/978-981-10-6116-5_11

traditional joint family system has been deteriorating fast and showing a nucleation of family system and changing value patterns among younger generations (Nayar 1999; Singh 1997). Therefore such strong filial obligations and mutual ties, the main feature of the traditional joint family, are weakening and have placed the older adults at economic and psychological stress (Sharma and Dak 1987). Kashyap (2004), reviewing various studies (Das and Bardis 1978; Gupta 1978; Kapur 1978; Jha 1985; Nair 1986; Rao and Rao 1982) argues that the modernization and subsequent social changes have affected the family structure, social roles, relationships and status of its members and its relationship with the kinship system.

Abuse includes physical assault, psychological aggression, violation of personal rights, sexual abuse, financial exploitation, neglect, and self-neglect. Such forms of abuses are particularly damaging to the older adults as most of the abusers are the victims' own adult offspring or trusted ones (Yan et al. 2015). Among these varied forms of elder abuse, verbal abuse has been identified as the most common type of mistreatment (Comijs et al. 1998; National Elder Abuse Incidence Study (NEAIS) 1998; Ogg and Bennett 1992; Yan and Tang 2001). Besides, physical abuse (Pillemer and Finkelhor 1988), financial abuse (Podnieks 1992) and neglect (Valentine and Cash 1986) have also been found to be common forms of elder abuse.

Elder abuse is a universally prevalent phenomenon and studies have shown that in the USA about 5–10% of people aged 65 or above have been abused by their care-givers on whom they depended for everyday activities (Fulmer et al. 2002; National Research Council 2003). As many countries are proceeding towards rapid ageing of the population, the number of elderly people who become potentially vulnerable to abuse is also expected to increase. This is particularly acute in Asia, whose population is ageing at an unprecedented pace. In, 2012, 11% of Asians were aged 60 or older, a figure expected to rise to 24% by 2050 (HelpAge International 2012). The greatest increase in the ageing population will occur in those over 75, from 15% in 2000 to 27% in 2050 (HelpAge International 2013). In this chapter we explore the different forms of emotional abuse as reported and experienced by the older adults both in home and institutional settings.

Contextualizing Elder Abuse and Mistreatment

Old age and care of older adults were never a problem for India when a value-based joint-family system prevailed and the older adults enjoyed support and comfort in the joint-family system (Chokkanathan and Lee 2005). Revering the older adults was an integral part of the culture (HelpAge India 2012). A significant proportion (84.4%) of the older adults in India lives or co-resides with their children (Rajan and Kumar 2003). But now this delicate dyadic care relationship between the care-receivers and the care-providers has greatly changed and is being renegotiated. The changing inter-generational relation has created a care difference or care-gap in the present-day care practices, both in the family care and in institutional old-age care. This care-gap is viewed differently by both the older adults who see it as a function of societal changes and their care-givers who are sandwiched between the care needs of two generations. Studies by HelpAge India (2012) have found that inadequate housing in metropolitan cities sometimes led to a lack of physical and emotional space or basic necessities that make the older parents shift to one corner of the house. Moreover, with increasing dual career households, the older adults who co-reside with family members are becoming more marginalized, isolated and insecure. In this context there is growing concern among older adults that they are more often being abused and neglected.

To understand the issue of elder abuse in any society it is necessary to understand the cultural background of that society because inter-personal and inter-generational relations are culturally constructed (Yan et al. 2015). Traditionally, in the Asian cultural context, adult children provide care, respect and financial support for their parents with reverence (Cheng and Chan 2006; Ng et al. 2000, 2002; Sung 2001). Earlier studies have shown that older adults continued to have high filial expectations of the younger generation and that young people still accepted these obligations for their parents (Lee and Sung 1997). However, recent findings show that young people are now likely to interpret filial duty differently from their parents, and this reciprocal filial duty depends on their future circumstances (Tsai et al. 2008). Tam and Neysmith (2006) in their qualitative study of home care-givers in China reported that disrespect is the main form of elder abuse in the Chinese community. Similarly, disrespect and lack of dignified living conditions are also considered as a major form of elder abuse by older Indians (HelpAge India 2012; Nagpaul 1998). Qualitative interviews with Chinese older adults affirmed that disrespect leads to unsettled feeling as expressed in phrases like *being ignored by children* and *behaving as if* [the elder person] *is the enemy* (Dong et al. 2010).

Similar observations have also been made in other Asian cultures. Chang and Moon (1997) found that older Koreans consider lack of respect and inappropriate care by the family members as major forms of elder abuse. They used examples such as 'failure to employ language that denotes respect', 'direct expression of disagreement with the mothers-in-law', from their studies to show the cultural perceptions of Korean elder abuse. The study of Arai (2006) in Japan showed that *blaming* by the young generation is the prominent form of abuse as older parents are blamed for whatever problems the adult children are facing. While the studies of Anme et al. (2006) stated that 17.9% of Japanese older adults living in agricultural villages reported abuse and 34.9% of family care-givers were engaged in potentially harmful behaviour against the older care recipients (Sasaki et al. 2007).

Notwithstanding the emerging consensus on perpetual elder abuse of Indian older adults (Prakash 2001; Vaswani 2001), the exact nature and magnitude is not well documented. The earlier studies showed that the existing evidence on elder abuse is sparse (Desai and Veedon 1993; Jamuna 1999) and even that was too largely based on qualitative studies (Jamuna 2003; Nagpaul 1998; Veedon 2001). Among 140 older adults studied, verbal abuse (80.7%) was the most frequent form of elder abuse followed by neglect (52.9%), financial abuse (37.1%), and physical abuse (22.9%) (Srinivas and Vijayalakshmi 2001 as cited in Chokkanathan and Lee 2005). While the studies of Chokkanathan and Lee (2005) revealed 14.1% prevalence of elder abuse in a sample of 400 community-dwelling Indians aged 55 or above, with chronic verbal abuse being the most common (10.8%) followed by financial abuse (5%), physical abuse (4.3%) and neglect (4.3%). In a more recent study of a representative household survey of 300 older Indians, Sebastian and Sekher (2011) argued that nearly half of the participants (49%) reported having experienced abuse or neglect from their family members. HelpAge India (2012) in a large-scale representative study of 5400 community-dwelling participants aged 60 or above found that 31% of older adults had been abused and 24% experienced abuse on a daily basis. But very few studies have focused on emotional neglect of older adults due to loneliness arising out of inter-generational relation conflict and gap. Apart from all these forms of elder abuse mentioned above, we made an attempt to bring out emotional and verbal forms of elder abuse in this study.

In the Asian cultural context, the issue of elder abuse is perceived as a private family matter (Dong et al. 2007; Yan and Tang 2001). Therefore, it is difficult to theoretically frame the nature and magnitude of elder abuse. The older adults predominantly perceived that confessing one's own abuse meant acknowledging to others that family members, especially adult children, were not fulfilling their filial obligations of respecting and caring for their elders (Cheng and Chan 2006). Disclosing family matters with others, particularly if the problem is one's own child's abusive behaviour, is perceived as shameful because such behaviour may be attributed to poor parenting (Lee and Eaton 2009). In Indian culture, acknowledging and reporting abuse are considered a taboo topic. A study in India found that 55% of older adults had not reported their abuses by their family member to anyone. Of these, 80% argued that they wanted to protect family honour and reputation (HelpAge International 2012 as cited in Yan et al. 2015). However, Desai and Raju (2000) attempted to explain elder abuse through different theoretical accounts: social isolation, stress, intra-individual dynamics, impairment dependence, negative attitudes towards older adults, and cycle of abuse of inter-generational transmission of violence. In this chapter an attempt has been made to explain the forms and issues of elder abuse through emotional abuse and neglect of older adults due to inter-generational conflict (see Fig. 11.1). Emotional abuse is an infliction of distress, pain or suffering through verbal or non-verbal acts. It also includes insults, threats, intimidation, humiliation and harassment (Singh et al. 2015).

Methodological Approach

The study applied a qualitative research design. The methods employed included in-depth interviews, the non-participant observation and field diary. A total of 116 in-depth interviews were conducted, out of which 42 were from Kolkata, 37 were from Delhi and 37 from Kerala. In this study, 79 in-depth interviews of older men and their care-givers were conducted in Kolkata and Delhi, both from homes and

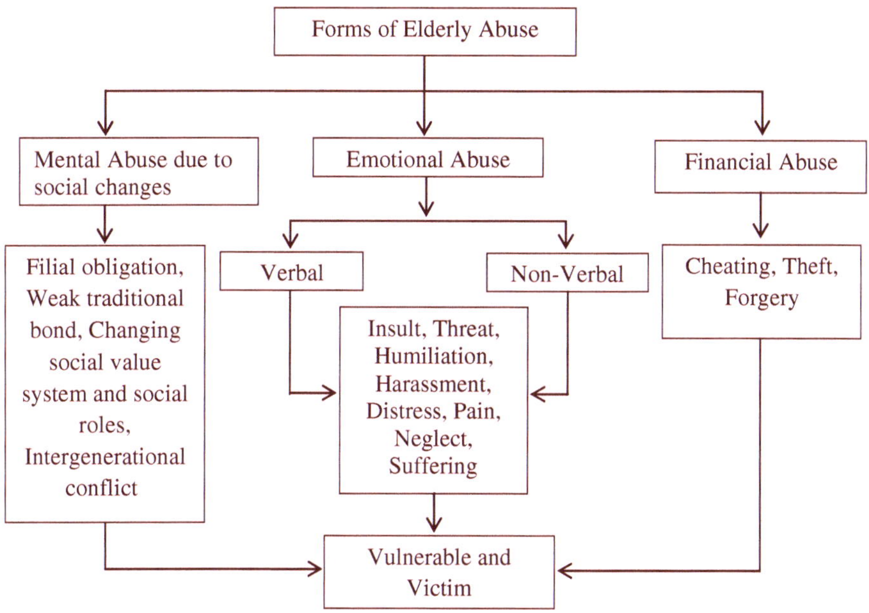

Fig. 11.1 Conceptual framework on forms of elder abuse. *Source* Authors

old-age institutions, to understand the forms of elder abuse and mistreatment in dyadic care relationships. Out of these 79 in-depth interviews, 47 were with older men and 32 were with their care-givers in Delhi and Kolkata. Out of these 47 older men, 25 were from Kolkata and 22 were from Delhi. Among the care-givers, 17 were from Kolkata and 15 were from Delhi. The main focus of the study in Kerala was to understand the place attachment and home-making process of older adults in various types of institutional care settings (old-age homes). There were a total of 37 in-depth interviews conducted at three old-age homes in Kerala. Each of these three old-age homes is used by older adults from different socio-economic classes. Among these 37 older adults, 23 were female and 14 were male participants. Besides, three more in-depth interviews were carried out among care-givers in each of the old-age homes. Participants were selected through purposive sampling, and interviews were conducted till the saturation of data. The data were collected in the months of June, July and August 2013 in Kolkata, and in the months of December 2013 and February and March 2014 in Delhi. On the other hand, the interviews in Kerala were conducted during the period from June to December 2015.

Participant Recruitment

In this study we have included older adults aged 60 and above who are dependent on the care-givers for day-to-day activities. The process of participant recruitment involved two stages: the first was to define the appropriate study population, and then to identify strategies for recruiting. The participants were accessed with prior permission in both the settings (i.e. in the homes as well as in the old-age homes). First, permission was taken from the authority of the institutions. Later we sought separate permission from participants, that is older adults and care-givers. We didn't recruit participants from among those who were sick and couldn't talk. Each participant was informed about the research before their participation and care was taken to get their consent. Photographs were taken and used only with the prior permission of participants. On the other hand, the older adults staying in homes were accessed through gatekeepers and were interviewed with prior permission of the family members. Still there were some constraints in accessing the older adults and their care-givers living in homes in Delhi. The constraints were mainly the age and gender of the first two authors who collected the data.

Interviews and Observation

The in-depth interviews were conducted in the institutional care centres (commonly known as old-age homes in India) and homes (Delhi and Kolkatta). An in-depth interview guide was used for the personal interview of the older men and their care-givers. The interview guide was prepared in English but was translated into the local language. There were few participants who preferred English over local language for the interviews. For older adults living at home, most of the interviews were conducted in park spaces where they go for meeting fellow older adults in the afternoons. Parks were selected so that interviewees could share their everyday life experiences away from their family members. The interviews for the institutional care-receivers were conducted in their personal room and in open spaces of the institutions.

Besides, much information was collected through observation as mere interviews cannot reveal the attitudes and the behaviour changes of the care-receivers as well as the care-givers in a particular setting. It also helped to cross-check the information shared by the participants in their interviews on particular issues such as behaviour towards each other. These observations were noted down in the field diary during the field study and incorporated while explaining the contexts.

Data Analysis

This study is based on the personal interviews data that were digitally recorded and transcribed in Bengali and in Hindi. All the transcribed data were translated into the English language. The interviews were analysed with the help of WeftQDA, a software package for qualitative data analysis. This software helped to develop codes for categories from the stories of the participants. Then each of the code family was described comparing different statements and quotes made by the participants. After that these descriptions were contextualized further with the observation data. Though the research was conducted through both the deductive and inductive methods, this particular study of perceptions and prevalence of elder abuse has come out as an inductive code. The results are analysed with the narratives of the care receivers and their emotional attachments and relationships with the people and place. For the interviews conducted in Kerala, we present the first results and impressions as the data was collected till December 2015.

Results and Discussion

The findings of the study indicate that the older adults report being ill-treated by their family members as well as by institutional care-providers. It has been observed that some of the older adults who were abused by their family members shifted to the institutions. Some continued to co-reside with children due to societal pressure and economic dependency. The older adults experienced mistreatment and verbal abuse as the prominent form of abuse. In this study, different forms of abuse have been discussed in two different settings; one is in the homes and the other is in institutions.

Multiplicities of Abuse

In a definition of the elder abuse it has been mentioned that one of the essential characteristics of elder abuse is the mistreatment by a trusted one in the family. In the interviews of the older adults, living in homes, they hesitantly narrated their bitter experiences. The most prominent form of emotional abuse is the use of foul language by the care-providers. **Verbal abuse**, disrespect and being threatened by the younger family members led to feelings of loss of status and security for the older adults. Alcohol addiction was mentioned as one of the precursors of violence. This may be seen below in the case of Kalam, where he tries to reason with his son about the ill-effects of alcohol on his already compromised state of health. The ensuing verbal abuse reflects the changing power relations where the son questions the ability of the father to stop the alcohol use. The consequence is older men, such as Kalam, retreat from such confrontations and refrain from engaging with the lives of their family members. Such disengagement then leads to loneliness and possible feelings of depression among older adults.

> When they become angry they abuse me; they use rough language. Once one of my sons was drunk and I asked him 'Why did you drink? We spend fifty thousand for your kidney

> recently and you are drinking why?' Then he replied, What the f*** you are saying? What can you do if I drink?' But he himself behaves like this. He drinks and takes name; 'What f*** you can do?' So I do not talk with him much. (Kalam, care-receiver at home, Kolkata)

Sebastian and Sekher (2011) state that older persons who are economically dependent on their children are more vulnerable to verbal abuse and neglect than those who are not. In this study some of the older adults mentioned that their family members are not compatible with them; therefore, they feel disrespected and neglected. Yan et al. (2015) report that inter-generational conflicts are likely to result from differing generational expectations and aspirations. Subhas perceives that loss of respect and abuse contribute to his suffering. This relative powerlessness against his son and daughter-in-law indicates the status of the older adult in the family. His use of the term 'today's daughter-in-laws' also shows that there is disconnect between what he expects from his daughter-in-law and the lived reality of him expressing his authority within the household. His coping mechanism then relies on withdrawing from any kind of communication and becoming more of a passive care receiver.

> No, no. Not at all. I am not only getting less care and respect but also I am suffering. I think the young generation feel pain for caring for the elderly. Today's daughters-in-law do not tolerate our words. If I say something strictly then they will fill their husbands' ears and then my sons will rebuke me `Why did you say like this to my wife?' Why should I say anything like this? Whatever they will provide I take and if they are not giving then I will not take; I will eat if they provide otherwise not eat; that's all. (Subhas, car-receiver in home, Kolkata).

Economic abuse in the forms theft of property, neglect in payment for living necessities and forcible transfer of assets was mentioned by most participants in our studies. The abuse by the family member, particularly by sons, was largely due to financial reasons such as distribution of property or sharing income in the family. We present here the case of Eliyamma who, though born into a rich family, has been left to fend for herself and has found shelter in a government old-age home. The case study highlights the experience of unmarried older women and their status in the families. Legal aid and protection also did not help Eliyamma get what is rightfully hers.

> Eliyamma Geevarghese, 82, is an unmarried woman; she is living in a government-run old-age home in Kerala. She had two brothers and two sisters but now she is the only one alive. She has several nephews and nieces. She was born into a rich family and got lot of ancestral property worth cores of rupees from her father. As she was unmarried she has suffered a lot. She was thrown out from her house in the middle of the night. Her brothers and their sons tried to grab her property. So she went to the Supreme Court in Delhi and won the case. The Court ordered that the property belongs to her and she must have the same standard of living as her brothers' families. She claimed that her nephews must pay her rent per month as per the Supreme Court verdict. However, one day when she visited her home, the extended family members didn't open the door. She is distressed because no-one from her family likes her presence; they don't even answer her phone calls.

Physical abuse was usually the result of inter-generational conflicts, and women were more vulnerable to such abuse than men in the study. In this study, participants revealed that they had been physical abused largely due to financial issues. This conflict could also be due to unmet care of the older adults and care-gap in their expectations. Such inter-generational conflicts sometimes lead to strong disputes and result in physical abuse of the older adults. The perpetrators of violence come largely from within the family and in most cases are children or spouses. The case study of Chempakam shows how physical violence is closely related to economic abuse.

> Chempakam is a 75-year-old woman who lives in Govt. Old Age Home in Kerala after she was mistreated in home by her son and daughter-in-law. She had handed over each and every property to her son and daughter-in-law. Her son also took money from her without her knowledge. She mourns that she has given everything to them and has nothing left for her own use. Once her son brought bottles of liquor though he already had enough *kallu* (liquor). Then he behaved badly, her daughter-in-law called the police. Police came to her house and enquired. Her son thought it was his mother who called the police. But she does not like him to be touched by police, to be maltreated. Then he threatened that '*if he is caught by police as informed by someone, he will come back and do his first murder. Isn't that me?*' Thus when there arose a quarrel, he pulled her hair and hit her. After the hit she cried out for help but nobody came from neighbourhood. Her family members did not tell her to go away but she herself decided to live in an institutional care centre largely due to the fear of being physically abused again.

Older women are the ones who are more vulnerable and suffer the larger share of abuse. The change in power relations at home upon the death of the husband makes the lives of the older women more precarious. They have less access to family property and legal aid. Those who do manage to take control of their assets are abused, cheated and abandoned by the family.

> My husband is a drunkard. I have suffered a lot. After drinking he used to attack me physically. He grabs me and throws me away. Not only my husband's attack but I had to suffer the assaults of his first son from his first marriage. One day he grabbed me and pushed me and hit me hard. (Saradha, care receiver in institution, Kerala).

Neglect and Mistreatment

Neglect is another prominent form of elder abuse in the contemporary changing family dynamics and deteriorating inter-generational relations. In this study the participants articulated how they were being neglected by their family members as well as by society. Neglect in the domestic setting comes in many forms: lack of communication, not providing for the necessities for the older adults, not providing food, not taking care of health problems or doctor visits. The care-gap is due to the unmet care expectations of the older adults.

> Whatever I say they might not like and the answers are given by the daughters-in-law not the son. So this becomes intolerable. And now what has happened is my sons have no say in anything, they support anything that their wives say. (Raghu, home care-receiver, Kolkatta).

In the institutional setting, older adults who expressed their opinions on negligence had to bear the consequences of complaining to the higher authorities. There were instances where the care-providers also articulated about the negligence of the older adults' care. It has been found that there is a close relation between negligence and loneliness; that is, the neglected care-receivers felt the loneliness in homes as well as in the institutions. The older men used words like *just passing my time, no hope, avoid me, not give attention* and *does not care* to express their discontents. In Delhi and Kolkata, the majority of care-receivers in the institutions perceived that they were dumped by the family members first then neglected and deserted in the institutions. They perceived that as they age and stay in the institutions they are less respected in society. They realized that when they were in homes people used to respect them. The case was different in Kerala, where older adults paid large amounts of money to stay in old-age home. This changed the power relations as they could demand the care they paid for. One has to also acknowledge that some older adults may not reveal the mistreatment due to societal norms and personal inhibitions of being the seen as the victim. Similarly, Nagpaul (1998) and Pablo and Braun (1998) report that older adults refuse to acknowledge such experiences of abuse because of personal disgrace, family honour and protecting family members' reputations (Gupta 2007).

Abuse and Push to Institutions or Independent Living

To avoid being abused and neglected by their family members, some of the older men preferred to leave the house and stay in a separate house with a domestic maid. In the Indian scenario the burden of care for older men is largely on the women in the household; in cases where the spouse is available, she is the main care-giver and after her it is the daughter-in-law. Incompatibility of co-residence with son and daughter-in-laws was seen as the major reason for inter-generational conflicts. Chokkanathan and Lee (2005) also found that inter-generational gap, adjustment problems of daughter-in-laws in a new environment, mother-in-laws' authoritative attitudes and sons' active or passive support of wives are cumulative reasons for elder abuse in the changing social and family system. To avoid such conflicts, many older adults then lived apart or moved into an institutional care home.

> Whatever I say they might not like and the answers are given by the daughters-in-law not the son. So this becomes intolerable. And now what has happened is my sons have no say in anything, they support anything that their wives say. But I think if we were living somewhere nearby this kind of thing might have happened. But since I do not live in their proximity I am saved from such abuses. This is the reason I stay away from them. (Manoj, care-receiver at home, Kolkata)

Other older adults stated that their family members, particular sons, were rude and abusive and they were not ready to spend on their fathers' medical and other aspects of care. So some older adults left the home and shifted to institutional care centres. Those who decided to live apart from their children or moved to care homes where they had to pay for the care were relatively well-off, most often with a pension by which they could afford the facilities.

> I became a burden for them and they asked me to get out of the house. I worked hard for my family. All my assets utilized for this family. Now I became a useless fellow for my family. They treated me as like their slave. They wanted me to become their slave'. (Parameswaran, care-receiver in institution, Kerala)

> They curse me, saying I should die. They say that we should die so that they don't have to spend so much of money on us. I had to get a pacemaker. What can I expect from them, they are educated, they are graduates; one of them has a MSc degree, what is the use of it? For whom did I spend lakhs of rupees? So I got disgusted and came here. I don't have a lack of money. By God's grace, there is no lack of house or car, but I am fed up. I have told my children that they can do anything but they should not tell me anything. (Soumen, care-receiver, Kolkatta)

There is a strong gender discrepancy in this movement to old age homes. Women who were not in paid employment or who were abandoned by their family members usually found themselves at the mercy of the government old-age homes. Conversely, men who had assets and pensions were in better position to choose a care home that matched their requirements.

Conclusion

In this chapter we have explored the various forms of abuse faced by older adults both at home and in institutional settings. Abuse of any kind on a vulnerable population is a matter of concern. As shown in this rich empirical material, the larger share of abuse takes places within domestic settings and the perpetrators are often family members. The powerlessness older adults and their inability to seek protection make elder abuse less visible for social intervention. The application of rigorous qualitative methods has enabled us to examine this sensitive issue in both the settings. The findings clearly reveal the need for government and community interventions to protect the rights of older adults. The gender dimension in these case studies further highlights the inequalities and discrimination women face in Indian society. Patriarchal norms and loss of rights gravely affect older women and make them more vulnerable to abuse. Based on our study, we recommend setting up helplines and elderly support groups which can aid in finding psychological and legal support for the older adults. Abuse in any form and at any age robs the victims of their dignity. It is one of the basic human rights we need to protect to promote the well-being of older adults.

Acknowledgements We would like to thank the participants of the project, older men, women and their care-providers, for their time and for sharing their views on care practices. We are also thankful to Shinjini, Rupsa and Dipankar for transcribing and translating the data. This chapter was written within the Indian-European research networking grant: Ageing and well-being in a globalising world (NWO 465-11-009), funded by NWO-ESRC-ICSSR. The participating institutions include the Institute for Social and Economic Change, Bangalore; the Centre for Development Studies, Trivandrum; the Population Research Centre, University of Groningen, The Netherlands; and the University of Southampton, United Kingdom.

References

Anme, Tokie, Mary McCall, and Toshio Tatara. 2006. An exploratory study of abuse among frail elders using services in a small village in Japan. *Journal of Elder Abuse & Neglect* 17: 1–20. doi:10.1300/J084v17n02_01.

Arai, Mizuho. 2006. Elder abuse in Japan. *Educational Gerontology* 32: 13–23. doi:10.1080/03601270500338567.

Chang, Janet, and Ailee Moon. 1997. Korean American elderly's knowledge and perceptions of elder abuse: A qualitative analysis of cultural factors. *Journal of Multicultural Social Work* 6: 139–154. doi:10.1300/J285v06n01_09.

Cheng, Sheung-Tak, and Alfred C.M. Chan. 2006. Filial piety and psychological well-being in well older Chinese. *The Journals of Gerontology Series B: Psychological Sciences and Social Sciences* 61: 262–269. doi:10.1093/geronb/61.5.P262.

Chokkanathan, Srinivasan, and Alex E.Y. Lee. 2005. Elder mistreatment in urban India: A community based study. *Journal of Elder Abuse & Neglect* 17: 45–61. doi:10.1300/J084v17n02_03.

Comijs, Hannie, Anne Margriet Pot, Johannes H. Smit, Lex M. Bouter, and Cees Jonker. 1998. Elder abuse in the community: Prevalence and consequences. *Journal of the American geriatrics Society* 46: 885–888. doi:10.1111/j.1532-5415.1998.tb02724.x.

Das, Man Singh, and Panos Demetrios Bardis (eds.). 1978. *The family in Asia*. New Delhi: Vikas Publishing House.

Desai, K.G., and Rosamma Veedon. 1993. Family dynamics and development programs. *Indian Journal of Social Work* 1: 132–138.

Desai, Murli, and S. Siva Raju. 2000. *Gerontological social work in India: Some issues & perspectives*. Delhi: B.R. Publishing Corporation.

Dong, Xin Qi, Melissa A. Simon, and Martin Gorbien. 2007. Elder abuse and neglect in an urban Chinese population. *Journal of Elder Abuse & Neglect* 19: 79–96. doi:10.1300/J084v19n03_05.

Dong, Xin Qi, E-Shien Chang, Esther Wong, Bernarda Wong, and Melissa A. Simon. 2010. How do US Chinese older adults view elder mistreatment? Findings from a community-based participatory research study. *Journal of Aging and Health* 23: 289–312. doi:10.1177/0898264310385931.

Fulmer, Terry, Gregory Paveza, and Lisa Guadagno. 2002. Elder abuse and neglect: Policy issues for two very different problems. *Public Policy & Aging Report* 12: 15–18. doi:10.1093/ppar/12.2.15.

Gupta, Giri Raj. 1978. The family in India: The joint family. In *The family in Asia*, ed. Man Singh Das, and Panos Demetrios Bardis, 72–88. New Delhi: Vikas Publishing House.

Gupta, Rashmi. 2007. Perceived caregiver burden scale for caregivers of the elderly in India. *Journal of Applied Gerontology* 26: 120–138. doi:10.1177/0733464806297685.

HelpAge International. 2012. Ageing in the twenty-first century: A celebration and a challenge. http://www.helpage.org/download/50af6e9c8f44b. Accessed 9 Jan 2016.

HelpAge International. 2013. Global AgeWatch Index 2013: Insight report, summary and methodology. http://www.helpage.org/global-agewatch/reports/global-agewatch-index-2013-insight-report-summary-and-methodology. Accessed 9 Jan 2016.

HelpAgeIndia. 2012. Report on elder abuse in India. http://www.globalaging.org/elderrights/world/2012/report%20on%20elder%20abuse%20in%20india.pdf. Accessed 9 Jan 2016.

Jamuna, Duvvuru. 1999. Elder care in India. Conflicting tradition and modernization. In *Aging societies in a new millennium. Proceedings of 3rd international conference of IAHSA*, Honolulu, Hawaii, 33–35.

Jamuna, Duvvuru. 2003. Issues of elder care and elder abuse in the Indian context. *Journal of Aging & Social Policy* 15: 125–142. doi:10.1300/J031v15n02_08.

Jha, Kali Nath. 1985. *Women towards modernization*. Patna: Janaki Prakashan.

Kapur, Promilla. 1978. The family in India: Women in modern India. In *The family in Asia*, ed. Man Singh Das, and Panos Demetrios Bardis, 72–88. New Delhi: Vikas Publishing House.

Kashyap, Lina. 2004. The impact of modernization on Indian families: The counselling challenge. *International Journal for the Advancement of Counselling* 26: 341–350. doi:10.1007/s10447-004-0169-7.

Lachs, Mark, and Karl Pillemer. 2004. Elder abuse. *Lancet* 364: 1192–1272. doi:10.1016/S0140-6736(04)17144-4.

Lee, Yoon-Ro, and Kyu-Taik Sung. 1997. Cultural differences in caregiving motivations for demented parents: Korean caregivers versus American caregivers. *International Journal of Aging and Human Development* 44: 115–127. doi:10.2190/BLY5-07M1-6VVY-XJMX.

Lee, Hee Yun, and Charissa K. Eaton. 2009. Financial abuse in elderly Korean immigrants: Mixed analysis of the role of culture on perception and help-seeking intention. *Journal of Gerontological Social Work* 52: 463–488. doi:10.1080/01634370902983138.

Nagpaul, Kalma. 1998. Elder abuse among Asian Indians: Traditional versus modern perspectives. *Journal of Elder Abuse & Neglect* 9: 77–92. doi:10.1300/J084v09n02_07.

Nair, Gopinath P.R. 1986. Kerala, India. In *Migration of Asian workers to the Arab world*, ed. Godfrey Guntallike, 19–55. Tokyo: United Nations University.

National Elder Abuse Incidence Study (NEAIS). 1998. The administration for children and families and the administration on aging in the U.S. Department of Health and Human Services. http://aoa.gov/AoA_Programs/Elder_Rights/Elder_Abuse/docs/ABuseReport_Full.pdf. Accessed 8 Jan 2016.

National Research Council. 2003. *Elder mistreatment: Abuse, neglect and exploitation in an aging America*. Washington, DC: The National Academic Press.

Nayar, P.K.B. 1999. Changing role of the family in the care of the elderly. In *Ageing in India*, ed. S.D. Gokhale, P.V. Ramamurti, Nirmala Pandey, and B.J. Pendse, 168–178. Bombay: Somaiya Pubs. Pvt. Ltd.

Ng, Sik Hung, Cynthia S.F. Loong, An Ping He, James H. Liu, and Ann Weatherall. 2000. Communication correlates of individualism and collectivism talk directed at one or more addressees in family conversations. *Journal of Language and Social Psychology* 19: 26–45. doi:10.1177/0261927X00019001002.

Ng, Anita Ching Ying, David R. Phillips, and William Keng-mun Lee. 2002. Persistence and challenges to filial piety and informal support of older persons in a modern Chinese society: A case study in TuenMun, Hong Kong. *Journal of Aging Studies* 16: 135–153. doi:10.1016/S0890-4065(02)00040-3.

Ogg, Jim, and Gerry Bennett. 1992. Elder abuse in Britain. *British Medical Journal* 305: 998–999.

Pablo, Sharon, and Kathryn L. Braun. 1998. Perceptions of elder abuse and neglect and help-seeking patterns among Filipino and Korean elderly women In Honolulu. *Journal of Elder Abuse & Neglect* 9: 63–76. doi:10.1300/J084v09n02_06.

Pillemer, Karl, and David Finkelhor. 1988. The prevalence of elder abuse: A random sample survey. *The Gerontologist* 28: 51–57. doi:10.1093/geront/28.1.51.

Podnieks, Elizabeth. 1992. National survey on abuse of the elderly in Canada. *Journal of Elder Abuse & Neglect* 4: 5–58. doi:10.1300/J084v04n01_02.

Prakash, Indira J. 2001. Elder abuse: Global response and Indian initiatives. *Indian Journal of Social Work* 62: 446–463.

Rajan, S.Irudaya, and Sanjay Kumar. 2003. Living arrangements among Indian elderly: Evidence from national family health survey. *Economic & Political Weekly* 38: 75–80.

Rao, V.V.Prakasa, and V. NandiniRao. 1982. *Marriage, the family and women in India*. New Delhi: Heritage Publishers.

Sachan, Shweta, and Sukhminder Kaur. 2014. Elderly abuse and neglect in India. *Asian Journal of Home Sciences* 9: 276–279. http://www.researchjournal.co.in/online/AJHS/AJHS%209(1)/9_276-279_A.pdf. Accessed 5 Jan 2016.

Sasaki, Megumi, Yumiko Arai, Keigo Kumamoto, Koji Abe, Asuna Arai, and Yoko Mizuno. 2007. Factors related to potentially harmful behaviors towards disabled older people by family caregivers in Japan. *International Journal of Geriatric Psychiatry* 22: 250–257. doi:10.1002/gps.1670.

Sebastian, Daliya, and T.V. Sekher. 2011. Extent and nature of elder abuse in Indian families: A study in Kerala. *HelpAge India-Research and Development Journal* 17: 20–28.

Sharma, Madan Lal, and T.M. Dak (eds.). 1987. *Aging in India: Challenge for the society*. New Delhi: Ajanta Publications.

Singh, Avtar, Akshi Duggal, Ravinder Singh, and Upmesh K. Talwar. 2015. Elderly abuse: Suffering in silence-who is listening? Some concern for social work practices. *International Journal of Health* 3: 14–19. doi:10.14419/ijh.v3i1.4497.

Singh, Yogendra. 1997. Changing trends in the Indian family and the adjustment of the aged. *R&D Journal* 3: 31–42.

Srinivas, S., and B. Vijayalakshmi. 2001. Abuse and neglect of elderly in families. *Indian Journal of Social Work* 62: 464–479.

Sung, Kyu-taik. 2001. Elder respect: Exploration of ideals and forms in East Asia. *Journal of Aging Studies* 15: 13–26. doi:10.1016/S0890-4065(00)00014-1.

Tam, Sandra, and Sheila Neysmith. 2006. Disrespect and isolation: Elder abuse in Chinese communities. *Canadian Journal on Aging* 25: 141–152. doi:10.1353/cja.2006.0043.

Tsai, Hsiu-Hsin, Mei-Hui Chen, and Yun-Fang Tsai. 2008. Perceptions of filial piety among Taiwanese university students. *Journal of Advanced Nursing* 63: 284–290. doi:10.1111/j.1365-2648.2008.04711.x.

Valentine, Deborah, and Tim Cash. 1986. A definitional discussion of elder maltreatment. *Journal of Gerontological Social Work* 9: 17–28. doi:10.1300/J083V09N03_03.

Vaswani, Thrity G. 2001. Family care of the elderly: Abuse, neglect and abandonment. *Indian Journal of Social Work* 62: 492–505.

Veedon, Rosamma. 2001. Elder abuse in the urban context. *Indian Journal of Social Work* 62: 480–491.

World Health Organisation (WHO). 2008. A global response to elder abuse and neglect: Building primary health care capacity to deal with the problem worldwide: Main report. http://apps.who.int/iris/bitstream/10665/43869/1/9789241563581_eng.pdf. Accessed 12 Feb 2014.

Yan, Elsie, and Catherine So-Kum Tang. 2001. Prevalence and psychological impact of Chinese elder abuse. *Journal of Interpersonal Violence* 16: 1158–1174. doi:10.1177/088626001016011004.

Yan, Elsie, Ko-Ling Chan, and Agnes Tiwari. 2015. A systematic review of prevalence and risk factors for elder abuse in Asia. *Trauma, Violence, & Abuse* 16: 199–219. doi:10.1177/1524838014555033.

Chapter 12
Elder Abuse in HIV-Positive People

Ashish Goel, Astha Koolwal and Nishesh Jain

Introduction

Abuse, neglect and mistreatment among older persons living with HIV/AIDS (PLHA) have never been systematically evaluated apart from sporadic case studies that have dotted the literature. While those with human immunodeficiency virus (HIV) and acquired immune deficiency syndrome (AIDS) carry significant levels of abuse and mistreatment, those PLHA in their silver years could often be victims of even greater neglect and mistreatment. There are umpteen reports that indicate that PLHA often live with lifelong neglect and abuse ('AIDS Policy Law' 1996; AIDS Alert 2005; Chilemba et al. 2014; Galvan et al. 2004; Gielen et al. 1997; Glémaud et al. 2014; Greenspan 1999; Mkandawire-Valhmu et al. 2013; Simoni and Ng 2002). When age is added to the HIV status of the individual, the chances of abuse increase greatly. It had long been held that HIV is a disease of younger people due to the sexual inactivity of the older people, but this myth was disproved when O'Neill (1992) reported seropositivity in a 73-year-old homosexually active bachelor who presented to the Geriatric Unit with confusion. Older persons were recognized as a distinct subset of those living with HIV when sporadic reports started appearing in published literature in the late 1980s and the early 1990s.

A. Goel
Department of Medicine, University College of Medical Sciences, University of Delhi, New Delhi, India
e-mail: ashgoe@gmail.com

A. Koolwal (✉)
Anti-retroviral Center, University of Medical Sciences, New Delhi, India
e-mail: koolestastha@gmail.com

N. Jain
Diabetes Endocrine and Metabolic Centre, University College of Medical Sciences, New Delhi, India
e-mail: dr.nisheshjain@gmail.com

Mala Kapur Shankardass and S. Irudaya Rajan (eds.), *Abuse and Neglect of the Elderly in India*, https://doi.org/10.1007/978-981-10-6116-5_12

While the physicians, geriatricians and public-health specialists were intrigued by the unique features in newly diagnosed older victims of a rapidly spreading global menace, the psycho-social plight of these individuals was often forgotten and their distinctive psychological needs in terms of care, counselling, inter-personal relations and changes therein were never considered.

Our extensive literature search did not reveal any focused study looking at the occurrence of abuse in older people living with HIV and AIDS.

For ease of understanding, the present chapter has been divided into four parts to address the issue of abuse in older people living with HIV and AIDS. These are the issues related to (a) abuse of older people; (b) abuse of people living with HIV and AIDS; (c) older people living with HIV and AIDS; and (d) abuse of older people living with HIV and AIDS.

Case Vignette

Savitri was a 70-year-old lady who lay in the heat of fever for four months, during which time she continued visiting the primary-care physician. She continued to tell her doctors and care-takers that she had a generalized weakness, which had progressed swiftly for the previous three months. She was usually accompanied by her care-giver from the old age home where she had been living for the past 15 years of her life, abandoned by her own children. She was reluctant to talk, avoided eye contact and was not able to answer most questions regarding her health or symptoms. The physician one day decided to interview her alone when her care-givers were not around. After several sessions it was learned that the poor lady had been assaulted and sexually violated multiple times by fellow residents at the old-age home. After further discussion with the care-giver, it was acknowledged that the lady had been diagnosed with HIV in the previous month and since then her health had been deteriorating, placing her in a vulnerable position for abuse.

Not long ago the typical face of HIV was envisioned as that of a young drug abuser involved in high-risk behaviour, but lately that face has been changing. It is not uncommon to find instead an old man who never suspected that he had HIV and does not know what high risk behaviour is.

Abuse of Older People

Definition of Abuse

The definition of abuse has been reviewed, refined, discussed and analysed but it still remains elusive. The National Research Council (NRC) report has defined elder mistreatment as ‘intentional actions that cause harm or create a serious risk of harm, whether or not intended, to a vulnerable elder by a caregiver or other person who is placed in a position of trust in any relationship to the elder or failure by a caregiver to satisfy the elder's basic needs or to protect the elder from harm'. The World Health Organization (WHO) defines abuse as 'single, or repeated acts, or lack of appropriate action, occurring within any relationship where there is an expectation of trust which causes harm or distress to an older person'. This important definition brings both acts of commission and omission into the purview of the definition of abuse.

Occurrence and Prevalence

Abuse is not just a five letter word but a way of life for those who suffer and for those who perpetrate the evil. It is not only wrong and unforgivable, but has also been likened to homicide. Despite these unmistakable facts, it remains excessively under-reported. Further, each victim is unique and every story is different. Every person who has been or is possibly being abused today may tell a different story with a unique set of circumstances surrounding him or her.

Elder abuse has been neglected over a very long time; just as much as has been the care of older people. It is noteworthy that in numerous households, abuse may be a way of life and may remain unreported forever. The statistics of abuse of older people may vary from country to country and the inherent cultural practices have a very crucial role to play in dictating the way elders are taken care of in a certain household. The same cultural practices may act as constraints preventing an older person from reporting the abuse or mistreatment that they are facing in the household.

The abuse rates may differ accordingly from country to country depending upon the cultural and traditional practices. It is believed that in ancient times, in South East Asian countries, the older adults had a revered position in the family and enjoyed supremacy, especially in terms of important decisions regarding land and property. They were respected and the family was likely controlled by them. Today, however, there has been a disintegration of the family structure, the families have become nuclear and older adults do not receive the same degree of reverence that their ancestors enjoyed in the past. In developing countries like Bangladesh, the prevalence of elder abuse as demonstrated in a recent study in 2009 by Tareque et al. (2015) was 62% in poor households and 6% in rich households among 896 older respondents. This prevalence however is much higher than that depicted by other studies in different countries. In a recent study published in 2015, Saikia et al. (2015) interviewed 331 community-dwelling older adults, using the Hwalek-Sengstock screening test. The results revealed a prevalence of 9.31%, which matches the result of studies in New York.

After segregating the mistreatment into different types of abuse, three studies consistently showed similar results in the same population. In a study on 4000 community-dwelling people in New York, the prevalence of elder abuse was found to be 7.6% (Burnes et al. 2015). In another survey by Laumann et al. (2008) in the same population, the prevalence was found to be 9%. A telephone survey conducted by Acierno et al. (2010) demonstrated that the rate of abuse was 10%. The survey was deficient on account of excluding people suffering from dementia, which has been found to be a risk factor for abuse. It has been stated by studies, however, that the risk of abuse has not been found to be increased by particular diseases. However, debilitated general condition has been associated with greater dependence and hence higher risk of abuse.

Abuse of People Living with HIV and AIDS

In the past, HIV was considered a disease of the young. The population usually suspected HIV was associated with those involved in high-risk activities such as intravenous drug abuse or commercial sex work. Also, older people have not been included in the demographic surveys for HIV. All these perceptions have led to under-estimation of the HIV-positive older people population.

With the advent of highly active anti-retroviral therapy (HAART) the life expectancy of HIV-positive people has increased (Muñoz et al. 1997). Clinically, HIV-positive older people have been found to have lesser counts of T-lymphocytes which are positive for cluster of differentiation-4 (CD4) than their younger counterparts, and they may have a higher response rate of conversion following initiation of drug therapy against the HIV virus. Justice et al. in the Veterans Ageing Cohort Study looked at HIV-infected and uninfected persons in infectious disease clinics and general medical clinics in 2006 and demonstrated that the mean CD4 cell count decreases with age after adjustment with all parameters (Justice et al. 2006).

With increasing age, the immunity and the ability to fight infection with the virus deteriorates, and both the T cell and the B lymphocyte cells in the blood are also depressed. The thymus gland atrophies and has a larger proportion of perivascular tissue that the stromal part which leads to decreased T cell maturation (Aw et al. 2007). According to Szabo et al. (1998) the formation of B-cells is reduced due to bone marrow suppression and reduced production of external stimulators like interleukin 7 (IL-7), and this can be restored after the external. Thegut found associated lymphoid tissue (GALT) is more likely affected in HIV and the virus replicates faster in the lymphoid cells in the intestine (Effros et al. 2008). In older people the gut lymphoid tissue undergoes age-related changes and is especially vulnerable.

The interest in older people living with HIV and AIDS has increased over the last few decades with the advent of highly active drugs that make life with HIV sustainable. The prevalence of HIV in the older people has been found to be between a third and a quarter of the prevalence in the 18–39 year age group. Eldred and West have reported that of all the HIV-positive reports in Canada, nearly 12% are people older than 50 years of age (Eldred and West 2005). In fact, Freeman and West in 2012 showed that, in Malawi, the prevalence of HIV positivity is almost twice as high among older people (50–64 years; 8.9%) than among their younger counterparts (15–49 years; 4.1%) (Freeman and Anglewicz 2012). The emergence of drugs which enhance sexual performance for older people and the ubiquitous use of HAART have increased the life expectancy of HIV/AIDS infected older adults. Ageing leads to multi-systemic changes which make an older adult dependent on their family members. In HIV there are additional physical and emotional complications that may cripple the normal life that an older adult leads. Okuno et al. (2014) studied the quality of life of 201 older people living in Brazil, based on the Brazilian quality of life instrument. The older people with HIV had the lowest scores in domains related to diagnosis disclosure/social stigma, sexual activity and financial support. The very fact of being diagnosed with a chronic and untreatable disease like HIV challenges mental steadiness and brings about numerous irreversible changes in a person's lifestyle. The unrelenting psychological stigma has a bearing on the response to the medication and may further prove detrimental (Gore-Felton and Koopman 2008). These individuals are also especially vulnerable to depression, loneliness and substance abuse.

Older people may not even suspect that they could be positive for HIV. Those who do not know that they have HIV are a more serious threat to themselves and society. This is in contrast to the higher level of awareness among younger people, who will often come and request an HIV test. The fact that the suspicion itself is a surprise poses a unique hurdle for the counsellor to surmount. The treatment counselling for these persons therefore becomes very crucial. Learning that they are HIV-positive brings a major life change and may revolutionize their family dynamics. It is sometimes difficult to identify a care-giver for these individuals, especially those who are living alone. In those who are living with their spouses/children, the care-giver can be easily identified and appropriate counselling be given. It has also been found that the risk of abuse is higher among those who live with their family/spouse, with the abuser being the care-giver on many occasions. The diagnosis of this chronic and long-term disease makes the patient particularly dependent on the care-giver for medications and for help, especially because of the rampant opportunistic diseases. In this vulnerable position, when the care-giver is aware of these circumstances, he/she and the other family members may take advantage and be likely to abuse the older adult. In a study in New York, which was done on 4000 older adults, it was found that abuse rates were higher among women and those belonging to lower social-economic backgrounds (Lachs and Pillemer 2015).

The response of therapy in older people has been studied with mixed results. The majority of the studies point out that the CD4 cell response/trend after treatment tends to be lukewarm among older people as compared to younger ones, and the viral load also remains higher. A few studies point out that the CD4 count is inversely related to age, stating that the functional component of the thymus is an important predictor of the response (Lederman et al. 2000; Manfredi and Chiodo 2000). A few studies however state that the viral and immunological response is better among older people which may be attributed to better adherence (Knobel et al. 2001; Paredes et al. 2000).

A significant problem in managing older patients in general and PLHA is the use of multiple drugs in view of numerous comorbid conditions. The presence of concurrent polypharmacy often confuses patients and care-givers, and frailty, dependence and disease may further complicate the diagnosis and management. The counsellor is in a special position to analyse the family dynamics of the patient and screen for any possible abuse.

Older People Living with HIV and AIDS

Dong and Simon (2014) recently elicited the risk factors associated with elder abuse and developed a vulnerability index to predict elder abuse for community-dwelling older people, taking into account several variables including socio-demographic, psychological and health-related factors. Age, race, sex and self-reported medical conditions were included among the parameters assessed. It has been shown earlier that comorbidities like diabetes, hypertension, CAD, fractures, stroke and cancer have been correlated to higher risk of elder abuse (Lachs and Pillemer 2015). Cognitive functioning and physical ability were also assessed in the study, and it was stipulated that poor cognition can be related to higher risk of elder abuse. The physical ability was also analysed through the Katz index, and the ability to perform daily life functions was assessed. Poor physical functioning in terms of difficulty with activities of daily living (ADL) has also been linked to high risk of elder abuse. The female sex, ages over 80, poor social network and more than three comorbidities have been linked to higher risk of elder abuse (Laumann et al. 2008).

There has been little focus on older people living in long-term care facilities. A few available studies indicate that abuse by staff members and fellow inmates is not unusual (McDonald et al. 2015). It is therefore implied that age and its associated senescence are only added factors which may facilitate abuse of older adults.

Implications

Emotional: The emotional impact of being abused by one's own children is enormous. One may find it especially difficult to cope in circumstances when trust is betrayed. Studies have shown that victims, especially females, are vulnerable to psychological disorders. Mouton et al. (2010) studied 93,676 women and assessed the prevalence of psychological problems in both abused and non-abused group. They found that the prevalence of depression, anxiety disorders, eating disorders etc. was more common in women who were abused either verbally, physically or financially.
Physical: With age there is an element of frailty or pre-frailty creeps in. Older people tend to have weaker bones and are prone to fractures. An insignificant fall or a previously trivial injury may cause disproportionate damage and may seem suspicious. A potential abuser may inflict serious injuries resulting in repeated fractures and other trauma to older people. Numerous fractures may lead to increased

dependence on the care-giver and may further increase the chances of future abuse. There may also be hastened onset of frailty, both physical as well as cognitive.

Compliance: Knobel et al. (2001) studied the response to HAART among older people, and demonstrated that they may have better adherence than middle-aged patients (Lederman et al. 2000). However, mixed results have been demonstrated in different studies. Older people may have adherence issues especially in cases where they are frail and potentially abused by the care-giver or someone at home. They may suffer from malnutrition and the care-giver may omit the administration of some medications. Neglect is worst for older people who are entirely dependent on their care-giver for daily needs such as medicines, food and ADL. The compliance issues may become important in these cases. It is also noteworthy that elevated anxiety levels are associated with decreased response rates to HIV drugs (Table 12.1).

Table 12.1 Indicators of elder abuse and the screening questions that have to be asked in case of suspicion

Abuse	Indicator	Question	Specific risk factor in HIV
Physical	Burns, bruises, cuts, finger-marks, poor hygiene, dehydration	Do you feel safe at home? Is there anyone in your family involved in substance abuse?	Malnourishment Opportunistic infections
Psychological	Changes in sleep/eating patterns, fear, confusion, reluctance to talk or make eye contact	Does anyone in your family force you to do something which you do not want to do?	Increased stress or anxiety levels Feeling of being a burden Malnourishment and inability to support oneself financially
Neglect	Decubitus ulcers, malnourishment, non-adherence	In case of an emergency, whom do you ask for help? Do you have anyone to go out with you for shopping or doctor's appointments?	Increased dependency for drugs and for doctor's visits
Financial	Inappropriate/erratic bank account with-drawls, missing property, lack of amenities even when the older person could afford them	Are your family dependent on you for their daily expenditures? Does anyone take your belongings without your permission?	
Sexual	Complains of being assaulted sexually, stained undergarments, inappropriate sexual behaviour	Do you feel safe at home? Does anyone touch you without your permission? Do you get time alone for yourself?	Malnourishment and increased dependency on the spouse/family member

Abuse of Older People Living with HIV and AIDS

During interaction with an older person living with HIV or AIDS, one needs to be sensitive to some suspicious signs or gestures. It is always crucial to talk to the older person and the care-giver alone and then together to assess the adherence and various issues that may need to be resolved, some of which may be unique for that particular older individual.

A simple evidence of abuse may be poor adherence or compliance to prescribed medications, especially when the adherence had been good in the past and there has been a recent decline in the same. Taking time to speak to the older person about any possible issues may assist in detecting the loopholes in care from the perspectives of both care-giver and health-care-giver.

There may be several problems in routine and casual communication, especially with older persons with dementia who would be deemed unreliable for answering questions without assistance from their care-giver. Even when the older person is able to articulate and answer reliably, they may prefer to hide their circumstances. This may preclude development of a good doctor–patient relationship which is essential before a frank and free disclosure or communication from the patient's side can be expected. Medical conditions could lead to situations where an individual becomes so frail and fragile that there may be grievous injuries in the face of seemingly trivial trauma and minimal impact. It is not uncommon among older persons to find injuries and fractures due to simple and minimal injury. Sometimes it becomes increasingly difficult to differentiate which injury was accidental and which one was intentionally inflicted. Conditions like osteopenia, osteoporosis, cancer and frailty which are very common in this age group make exact characterization not only difficult but often impossible even to the trained eye. It has been seen that long bone fractures are very common in bedridden osteopenic patients. Fractures of the small bones of the hand are however found more commonly in cases of abuse (Burnes et al. 2015). Indeed, Connolly et al. (1995) reported the case of an elderly lady in 1995, where the authors found it extremely challenging to differentiate a case of possible abuse from a simple osteopenic fracture. There may also be numerous cultural and traditional differences involved in evaluating an individual for abuse. In cases of verbal abuse the cultural differences come into play when certain issues may or may not be regarded as potential abuse by the older people.

Emotional mistreatment may vary from country to country as expectations of the older people may vary. In Asian countries, the element of 'disrespect' is very prominent. The feeling of being disrespected may change emotional dynamics and make older people feel neglected, which is a feeling which may vary in intensity in different parts of the world.

A few of the older adults develop certain strategies called coping strategies, through which they acclimatize to the abusive environment in which they live. A few of them still keep hope and try to find peace in the tumultuous environment they inhabit. A few dismiss their domestic issues as an essential component of their life and try to distance themselves from these issues as much as possible. Abusive children most likely have substance abuse issues and their older parents are more prone to get affected with chronic pain and depression.

Conclusion

Though the first case of abuse was reported in the late twentieth century, it is a fact that older people have been abused for a long time. The definition, demography and prevalence of elder abuse have seen rapid changes and these have been reasonably defined in the last century. However, the risk of abuse in chronic debilitating diseases like HIV has not been described formally. It is not difficult to understand that the occurrence of abuse would certainly increase greatly in such situations and would acquire unique and specific issues not seen in other scenarios. The situation is not only further aggravated but also complicated due to the strange stigma surrounding a disease like the HIV-AIDS pandemic. We need to recognize the changing face of HIV and work on the different paradigms which may change for an HIV patient when entering his or her silver years.

References

Acierno, Ron, Melba A. Hernandez, Ananda B. Amstadter, Heidi S. Resnick, Kenneth Steve, Wendy Muzzy, and Dean G. Kilpatrick. 2010. Prevalence and correlates of emotional, physical, sexual, and financial abuse and potential neglect in the United States: The National Elder Mistreatment Study. *American Journal of Public Health* 100 (2): 292–297. doi:10.2105/AJPH.2009.163089.

AIDS Alert. 2005. Study examines impact of abuse on HIV patients. Emotional abuse is significant marker. *AIDS Alert* 20 (5): 55–57. http://www.ncbi.nlm.nih.gov/pubmed/15912616.

Aw, Danielle, Alberto B. Silva, and Donald B. Palmer. 2007. Immunosenescence: Emerging challenges for an ageing population. *Immunology* 120 (4): 435–446. doi:10.1111/j.1365-2567.2007.02555.x.

Burnes, David, Karl Pillemer, Paul L. Caccamise, Art Mason, Charles R. Henderson Jr., Jacquelin Berman, Ann Marie Cook, et al. 2015. Prevalence of and risk factors for elder abuse and neglect in the community: A population-based study. *Journal of the American Geriatrics Society* 63 (9): 1906–1912. doi:10.1111/jgs.13601.

Chilemba, Winnie, Neltjie van Wyk, and Ronell Leech. 2014. Experiences of emotional abuse among women living with HIV and AIDS in Malawi. *Health Care for Women International* 35 (7–9): 57–743. doi:10.1080/07399332.2014.926901.

Connolly, V., A.A. McConnell, and G. McGarrity. 1995. Battered granny or spontaneous fractures? A legal dilemma. *Postgraduate Medical Journal* 71 (840): 630–632.

Dong, XinQi, and Melissa A. Simon. 2014. Vulnerability risk index profile for elder abuse in community-dwelling population. *Journal of the American Geriatrics Society* 62 (1): 10–15. http://www.ncbi.nlm.nih.gov/pmc/articles/PMC4155408/.

Effros, Rita B., Courtney V. Fletcher, Kelly Gebo, Jeffrey B. Halter, William R. Hazzard, Frances McFarland Horne, Robin E. Huebner, et al. 2008. Aging and infectious diseases: Workshop on HIV infection and aging: What is known and future research directions. *Clinical Infectious Diseases: An Official Publication of the Infectious Diseases Society of America*. doi:10.1086/590150.

Eldred, Susan, and Lynne West. 2005. HIV prevalence in older adults. *The Canadian Nurse* 101 (9): 20–23. http://www.ncbi.nlm.nih.gov/pubmed/16350736.

Freeman, E., and P. Anglewicz. 2012. HIV prevalence and sexual behaviour at older ages in rural Malawi. *International Journal of STD and AIDS* 23 (7): 96–490. doi:10.1258/ijsa.2011.011340.

Galvan, Frank H., Rebecca Collins, David E. Kanouse, M. Audrey Burnam, Susan M. Paddock, Robin Beckman, and Steve R. Mitchell. 2004. Abuse in the close relationships of people with HIV. *AIDS and Behavior* 8 (4): 51–441. doi:10.1007/s10461-004-7328-z.

Gielen, A.C., P. O'Campo, R.R. Faden, and A. Eke. 1997. Women's disclosure of HIV status: Experiences of mistreatment and violence in an urban setting. *Women and Health* 25 (3): 19–31. doi:10.1300/J013v25n03_02.

Glémaud, Myriam, Lourdes Illa, Marisa Echenique, Victoria Bustamente-Avellaneda, Shirley Gazabon, Olga Villar-Loubet, Alan Rodriguez, et al. 2014. Abuse and mental health concerns among HIV-infected Haitian women living in the United States. *The Journal of the Association of Nurses in AIDS Care: JANAC* 25 (1 Suppl): S62–S69. doi:10.1016/j.jana.2013.06.005.

Gore-Felton, Cheryl, and Cheryl Koopman. 2008. Behavioral mediation of the relationship between psychosocial factors and HIV disease progression. *Psychosomatic Medicine* 70 (5): 569–574. doi:10.1097/PSY.0b013e318177353e.

Greenspan, J. 1999. Stop the abuse of HIV+ women prisoners at Chowchilla. *World* (97): 4. http://www.ncbi.nlm.nih.gov/pubmed/11366855.

Justice, Amy C., Elizabeth Dombrowski, Joseph Conigliaro, Shawn L. Fultz, Deborah Gibson, Tamra Madenwald, Joseph Goulet, et al. 2006. Veterans Aging Cohort Study (VACS). *Medical Care* 44 (Suppl 2): S13–S24. doi:10.1097/01.mlr.0000223741.02074.66.

Knobel, H., A. Guelar, G. Valldecillo, A. Carmona, A. Gonzalez, J.L. Lopez-Colomes, P. Saballs, J.L. Gimeno, and A. Diez. 2001. *Response to highly active antiretroviral therapy in HIV-infected patients aged 60 years or older after 24 months follow-up*. London: AIDS.

Lachs, Mark S., and Karl A. Pillemer. 2015. Elder abuse. *New England Journal of Medicine* 373 (20): 1947–1956. doi:10.1056/NEJMra1404688.

Laumann, Edward O., Sara A. Leitsch, and Linda J. Waite. 2008. Elder mistreatment in the United States: Prevalence estimates from a Nationally Representative Study. *The Journals of Gerontology* 63 (4): S248–S254.

Lederman, M.M., R. McKinnis, D. Kelleher, A. Cutrell, J. Mellors, M. Neisler, E. Cooney, et al. 2000. Cellular restoration in HIV infected persons treated with Abacavir and a protease inhibitor: Age inversely predicts naive CD4 cell count increase. *AIDS* 14 (17): 2635–2642.

McDonald, Lynn, Christine Sheppard, Sander L. Hitzig, Tal Spalter, Avantika Mathur, and Jason Singh Mukhi. 2015. Resident-to-resident abuse: A scoping review. *Canadian Journal on Aging = La Revue Canadienne Du Vieillissement* 34 (2): 215–236. doi:10.1017/S0714980815000094.

Manfredi, R., and F. Chiodo. 2000. *A case-control study of virological and immunological effects of highly active antiretroviral therapy in HIV-infected patients with advanced age*. London: AIDS.

Mkandawire-Valhmu, Lucy, Patricia E. Stevens, Peninnah M. Kako, and Anne Dressel. 2013. Safety planning in focus groups of Malawian women living with HIV: Helping each other deal with violence and abuse. *Violence Against Women* 19 (11): 49–1331. doi:10.1177/1077801213514859.

Mouton, Charles P., Rebecca J. Rodabough, Susan L.D. Rovi, Robert G. Brzyski, and David A. Katerndahl. 2010. Psychosocial effects of physical and verbal abuse in postmenopausal women. *Annals of Family Medicine* 8 (3): 206–213. doi:10.1370/afm.1095.

Muñoz, Alvaro, Caroline A. Sabin, and Andrew N. Phillips. 1997. The incubation period of AIDS. *AIDS* 11 (12): S69–S76.

Okuno, Meiry Fernanda, Alexandre Cavallieri Pinto, Leticia Meazzini Gomes, Gerson Scherrer Junior, Domingos Belasco Junior, and Angelica Goncalves Silva Belasco. 2014. Quality of life in elderly patients living with HIV/AIDS. *Cadernos de saude publica* 30 (7): 1551–1559.

O'Neill, D. 1992. HIV and the older people. *Journal of the Royal Society of Medicine*.

Paredes, R., A. Mocroft, O. Kirk, A. Lazzarin, S.E. Barton, J. van Lunzen, T.L. Katzenstein, F. Antunes, J.D. Lundgren, and B. Clotet. 2000. Predictors of virological success and ensuing failure in HIV-positive patients starting highly active antiretroviral therapy in Europe: Results from the EuroSIDA study. *Archives of Internal Medicine* 160 (8): 1123–1132.

Saikia, Anku Moni, Neelakshi Mahanta, Ajaya Mahanta, Ashok Jyoti Deka, and Arupjyoti Kakati. 2015. Prevalence and risk factors of abuse among community dwelling elderly of Guwahati City, Assam. *Indian Journal of Community Medicine: Official Publication of Indian Association of Preventive & Social Medicine* 40 (4): 279–281. doi:10.4103/0970-0218.164406.

Simoni, Jane M., and Ming T. Ng. 2002. Abuse, health locus of control, and perceived health among HIV-positive women. *Health Psychology: Official Journal of the Division of Health Psychology, American Psychological Association* 21 (1): 89–93. http://www.ncbi.nlm.nih.gov/pubmed/11846349.

Szabo, P., K. Zhao, I. Kirman, J. Le Maoult, R. Dyall, W. Cruikshank, and M.E. Weksler. 1998. Maturation of B cell precursors is impaired in thymic-deprived nude and old mice. *Journal of Immunology* 161 (5): 2248–2253.

Tareque, Md Ismail, Towfiqua Mahfuza Islam, Atsushi Koshio, and Kazuo Kawahara. 2015. Economic well-being and elder abuse in Rajshahi District of Bangladesh. *Research on Aging* 37 (2): 200–224. doi:10.1177/0164027514527833.

Two HIV-positive lawyers file suits alleging mistreatment. 1996. *AIDS Policy & Law* 11 (11): 4. http://www.ncbi.nlm.nih.gov/pubmed/11363562.

Chapter 13
Creating Conducive Spaces for Elder Abuse Interventions: Perceptions of Health-Care Professionals

Asha Banu Soletti and Pankhuri Bhatia

Elder abuse has devastating consequences for older people as it affects their health, mental health and well-being. The research data and media portray the increase of instances of elder abuse in the country. Yet there is a lack of trained professionals to identify and intervene in such scenarios. In India, there is no specially trained workforce in the field of gerontology. There are a few courses that attempt to train people to work with the elderly at various levels. However, these does not cover the gamut of issues surrounding seniors' lives. In the context of paucity of specific services for elder care, and particularly to address elder abuse, it is crucial to identify ways and means to train the existing professionals (like counsellors and doctors) who meet older people in different contexts. In order to initiate this process, it is essential to understand professionals' perceptions and awareness of elder abuse and the role they currently play to mitigate it.

This chapter draws from the qualitative data that have been collected from professionals located in Mumbai. The professionals included counsellors, social workers and doctors. In-depth interviews captured data on how elder abuse is viewed and addressed through the lens of the health-care workforce. The findings illustrate the need to really invest in capacity building of existing professionals regarding identification, screening and referral of cases of elder abuse. The chapter concludes with recommendations and strategies to build mechanisms that identify, intervene and suggest, creating conducive and approachable spaces within the existing system for elderly people in crisis.

A.B. Soletti (✉)
Centre for Health and Mental Health, School of Social Work, Tata Institute of Social Sciences, Deonar, Mumbai, India
e-mail: ashabanu@tiss.edu

P. Bhatia
Centre on Death Penalty, National Law University, Delhi, India
e-mail: pankhbhatia@yahoo.com

Mala Kapur Shankardass and S. Irudaya Rajan (eds.), *Abuse and Neglect of the Elderly in India*, https://doi.org/10.1007/978-981-10-6116-5_13

Introduction and a Brief Background

The world's population is continually greying and every country in the world is experiencing population ageing. According to data from World Population Prospects: the 2015 Revision (United Nations 2015), the number of older persons (those aged 60 and above) has increased substantially in recent years in most countries and regions, and that growth is projected to accelerate in the coming decades. Between 2015 and 2030, the number of people in the world aged 60 years and above is projected to grow by 56%, from 901 million to 1.4 billion; and by 2050, the global population of older persons is projected to be more than double its size in 2015, reaching nearly 2.1 billion. The global and country-wise data indicate the increase in the number of elderly; however, the policy, programmes and infrastructure in developing countries do not seem to acknowledge this transition. Ageing in developing countries that are undergoing a rapid social change, such as urbanization or industrialization, makes the elderly more lonely, vulnerable and dependent. In the neo-liberal context, the needs and challenges experienced by the elderly are very diverse depending on their social position. The inter-sectionalities of vulnerability have been studied and understood by researchers, but there are no subsequent programmes or policies that have been evolved to address these issues adequately. The existing programmes do not capture the diverse needs and the heterogeneity of this cohort. The lack of social measures and created dependency of older people increases the incidence of elder abuse.

A definition developed by the UK's Action on Elder Abuse and adopted by the International Network for the Prevention of Elder Abuse (INPEA) states: 'Elder abuse is a single, or repeated act, or lack of appropriate action, occurring within any relationship where there is an expectation of trust which causes harm or distress to an older person'. The usual categories that are articulated in the documents and reports are physical abuse, psychological abuse, neglect, financial abuse and sexual abuse. There is considerable variation in how elder abuse is articulated there are diverse theories about the phenomenon. Lack of consensus on what constitutes abuse is the most significant impediment to the development of an adequate knowledge-base on elder abuse.

Though elder abuse is not a new phenomenon, around the world there is currently increased attention and emphasis on addressing it. There are sufficient data to indicate the prevalence, patterns and contexts in which elder abuse occurs. Abuse is viewed as a public-health problem in many countries and governmental agencies and non-governmental organizations (NGOs) are attempting to tackle it through research, prevention and intervention strategies.

Silent But Loud: Elder Abuse on the Rise

> I have three sons, they all come back home drunk. They do not bother to give me money, but they expect that the food be kept ready when they come back home. If the food is not ready, they use all filthy words on me and even have pushed me physically in rage. I suffer

> silently. They have not realized that I am growing old and becoming fragile. I have to spend my rest of the days taking care of them and receiving only hurt from them. I do not have any other place to go. Hope they will realize my value when I am gone. (72-year-old mother in a local church)

> I am on the street now; I walked out because they did not want me with them. Every day there was some argument on some petty matters. Neither they are in peace nor am I. It is better to live on the street begging to strangers, rather than begging for attention from your own blood, because the latter hurts more. (Senior citizen on the street, age unknown)

> A young boy molested me on the street, but no-one believed my words. They thought I am imagining it and that it can never happen. I was physically and emotionally hurt. I thought it's better to suffer alone than sharing it with people who did not take my words seriously. Rather than being supportive, they ridiculed me. (66-year-old woman in the neighbourhood)

A nationwide study conducted by HelpAge India (an NGO which fights isolation, poverty and neglect of the elderly) concluded that elder abuse in India is on the rise. The report revealed that at least 31% of elders suffer abuse and 24% face abuse almost daily. Elder abuse has serious physical and mental health implications. It also affects the overall well-being of the elderly.

It has been noted through research by different scholars that elder abuse is associated with shame, guilt and distress resulting in increased mortality amongst victims. Despite the research and evidence on instances of elder abuse, discovering its true prevalence is challenging. A report on violence and health by the World Health Organization (2002) has captured the ramifications, contexts and prevalence of elder abuse around the world. The examples of elder abuse range from outright physical assault of old people in modernized cultures that have been sadly acculturated to so-called traditional forms of family violence, to the systematic ostracization of tribal elders by the community in some less developed countries as a form of scapegoating. The establishment of the INPEA in 1997, with representation from countries around the world, indicates increasing international concern about elder abuse (Lachs 2004).

The literature on elder abuse is woven around certain broad themes; prevalence of elder abuse, types of elder abuse, the main perpetrator, and the barriers to seeking help. There are many gaps that need to be examined and understood in this field. In India, elder abuse has been reportedly experienced by more than 20% of the elderly with the most common forms experienced being disrespect and neglect. Most of the elderly are experiencing it almost on an everyday basis. The HelpAge India report revealed that at least 31% of elders face abuse and 24% face abuse almost daily. Sons emerged as the main perpetrators with 56% of the cases, whereas in 26% of the cases daughters-in-law were the main perpetrators. Of the elderly, 45% have remained silent and haven't reported the incident. Some 23% did not know how to deal with the problem, so they refrained from reporting it. The survey cited emotional dependence, changing ethos and economic dependence as the veiled reasons for the abuse. According to the same survey, 55% were cases of disrespect, 39% verbal abuse, and 23% economic exploitation. The extent of reporting of abuse is low and this can be attributed to maintaining confidentiality of a family matter and fear of retaliation (HAI 2013). Disrespect, neglect and verbal abuse are the three major forms of abuse as understood by the elderly themselves, and the same are the most prevalent forms as reported by most. The studies indicate that in the Indian context the primary perpetrators of abuse among elderly people are the sons and daughters-in-law.

A review of the few studies that have focused on elder abuse indicates that the most likely victim of elder abuse is a female of very advanced age, role-less, functionally impaired, lonely and living at home with someone, primarily their adult child, spouse or other relatives. Studies in India (Raju 2002) indicate that more women than men complain of maltreatment in terms of both physical and verbal abuse. The prevalent patterns of elder abuse include mainly psychological abuse in terms of verbal assaults, threats and fear of isolation, physical violence and financial exploitation. The health profile of the elderly victims indicates that a person suffering from physical or mental impairment and dependent on care-takers for most of his or her daily needs is likely to be the victim of elder abuse. A study was conducted by Skirbekk and James, published in 2014 using the data collected by a 2011 national-level survey conducted across seven states in India by the United Nations Population Fund (UNFPA): 'Building Knowledge-Base on Population Ageing in India'. The states surveyed were Kerala and Tamil Nadu from the South, Maharashtra from the West, Orissa and West Bengal from the East and Punjab and Himachal Pradesh from the North. The findings indicated that 11% of elderly people aged 60 and above have experienced at least one type of elder abuse (physical 5.3%, verbal 10.2%, economic 5.4%, disrespect 6%, neglect 5.2%). The most common perpetrator was found to be the son, who is reported to be responsible for the abuse among 41% of male victims and 43% of female victims. In India, sons are expected to take care of their aged parents. The major reasons for such abusive behaviour include: lack of adjustment, economic dependence of the abused, and increasing longevity of the old. Family ties in India and other South Asian countries are strong and an overwhelming majority of the old live with their family members. However, the increasing number of older persons, the longevity of life and the prolonged dependency escalates the instances of elder abuse. The only conclusion that can be drawn from the available literature is that abuse of elderly people is increasing. The review of various studies has indicated that psychological abuse in terms of verbal assault, threats and fear of isolation/abandonment is more common than other forms of abuse.

Abuse and neglect often trigger a downward spiral, which overturns an otherwise productive, self-sufficient older person's life (Breaux and Hatch 2003). It is hard to establish exact data on how many elderly people are being abused in a domestic setting because of the sensitive nature of the problem. Particularly problems in this respect arise from there being no uniform definition addressing both abuse and neglect, and no comprehensive, effective reporting system in place. In Asia, studies by researchers in China (Hong Kong SAR), India, Japan and the Republic of Korea have drawn attention to the problem of elder abuse, but no official action, in terms of policies or programme development, has followed so far (WHO 2002).

Despite the fact that there are many reports and studies on elder abuse, it can be mentioned that the research and understanding of it are still at a nascent stage. The available data are limited, inconsistent, uncomprehensive and lack generalizability.

Barriers to Older People Reporting Abuse

Home is considered as a private domain and thus its matters are not expected to be discussed in public.

Senior citizens consider it particularly taboo to discuss private matters in public. Elderly people prefer to suffer in private rather than suffer the shame of reporting that their children are abusing them. There is a strong stigma associated with this phenomenon. Seniors are also reluctant to expose their children. In addition to this, there is often a lack of awareness about what constitutes elder abuse, and lack of knowledge about seniors' rights and resources.

Diminished cognitive capacity related to old age, and mental or physical disability restrict the mobility of elderly to a great extent, and thus they are dependent on others to take them to any facility, even a religious space that they are familiar with. Roads and public spaces in India are not so senior-friendly, and often do not permit elderly people to access facilities when they want to. Dependence on their children does not allow victims to talk about abuse.

Elderly women fear alienation and social isolation. There is a constant need to preserve a family relationship. These women do not come out in the open to report abuse because of the associated stigma and shame. The culture of any society also imposes barriers to seeking services. There may be a lack of options or access to services. Women's status and position in society lead to the formation of a gender role which is itself abusive in a way. All these factors prevent individuals, professionals and communities from recognizing, disclosing and responding appropriately to violence against elderly women (Bagshaw 2011).

Retirement is usually considered the beginning of old age. It is followed by lack of work which further leads to paucity of economic resources; there can be a lack of pension or other independent and ongoing sources of income. Health during this age is a matter of luck, and to top it all there is a general neglect of this population by the social services. All these factors combined can lead to entrapment of the victim in an abusive relationship (Buchbinder and Winterstein 2003). Women often do not share their stories to family members, friends and professionals, and hesitate to alert them to their plight. This appears to be the essence of survival and ageing in a violent environment. The literature devotes much attention to the unique needs of older, battered women, such as their need for sustenance, a safe place, intense counselling and treatment, acquisition of relevant information, assistance from health-care professionals, and social and emotional support (Vinton 1992, 2001).

According to HelpAge India's 2013 report, the police helpline is a widely known mechanism for redress among 70% of the elderly. Very few people were aware of the tribunal under the Maintenance and Welfare of Parents and Senior Citizens Act,

2007. Other family violence issues, such as domestic violence and child abuse, have proved that abuse, neglect and exploitation require a multifaceted solution, including approaches from public health, social services, and law enforcement. But while these other types of abuse have been recognized and received attention, elder abuse has not. Perhaps the greatest barrier to addressing elder abuse is its relatively low profile among the general public, the health-care community, and law enforcement compared to other issues such as child abuse and domestic violence. At home, families are reluctant to report it, while in institutional settings, other residents and employees are slow to report. Fear of increased abuse or retaliation, social stigma, or lack of whistle-blower protections are some of the reasons for lack of reporting. Even when reports are made, the response is sometimes inadequate and inappropriate.

There are few studies that have attempted to capture the professionals' views about elderly people not reporting abuse. In a study conducted in the USA, the reasons for lack of reporting have been identified by community health-care providers and include clinicians' reluctance to acknowledge abuse, lack of protocol to identify abuse, fear of liability and limited number of services available to address abuse (Schemeidel et al. 2012). Reasons for lack of case-detection decisions by health-care professionals include lack of knowledge about elder abuse, its prevalence, signs and symptoms, risk factors, and information about perpetrators (Schemeidel et al. 2012).

Creating Conducive Spaces

The literature review clearly brings out the prevalence and types of elder abuse. An increasing amount of research pays attention to the type of elder abuse and documents the same through quantitative and qualitative inquiries. Although these have brought to light the increase in elder abuse cases and their various types, adequate measures have not been taken to address this persisting problem. James and Graycar (2000) have argued that inconsistencies in methodologies and incomparability of studies render it difficult to come to conclusions about the full extent of abuse of older people. Apart from these, there are other factors which lead to gross miscalculations. Such factors include under-reporting of cases of abuse by the elderly, older people not being able to label or recognize certain behaviour as abusive, fear of talking about abuse in public, and the invisibility of certain types of abuse because of cultural sanctioning (Tam and Neysmith cited in Bagshaw et al. 2013). The problem is greatly escalated because of the paucity of courses creating trained professionals in the field of gerontology.

Thus, it is crucial to identify the ways and means of training the existing professionals like counsellors and doctors who meet elderly people face to face in different contexts. In order to initiate this process, it is essential to understand professionals' perceptions and awareness of elder abuse, and to grasp the role they currently play in mitigating it.

In India, there has so far been no focused study to understand the intervention mechanisms in the health-care system. This chapter is a first step towards exploring spaces within the health-care system. Literature from other countries has spoken about integration of elder-abuse services within the primary health-care level as older adults visit health facilities periodically. There are also comprehensive assessment tools that certain countries have evolved to assess elder abuse, and documented intervention plans for diverse contexts. This chapter's inquiry is exploratory in the Indian context, but definitely a first step towards achieving preparedness among health-care practitioners for addressing concerns around elder abuse.

Methods

The study used a qualitative methodology and the themes have evolved out of the unstructured in-depth interviews that were conducted with health-care workers, doctors, social workers and counsellors in the city of Mumbai. A total of nine participants were interviewed. Three counsellors, running their private practices, three social workers located in a public hospital and three general physicians from private clinics were interviewed. The interviews lasted between 20 and 40 min, and transcribed verbatim. Using a multi-step process of thematic analysis, the core themes that represent the perceptions of nurses, physicians and social workers were identified. The areas covered in the interviews were: understanding and perceptions of professionals about elder abuse, interventions, barriers and challenges to identification and intervention. Informed consent was sought from each participant explaining the nature of the study. Each participant willingly participated and shared their understanding on the issue. The content of the interviews was critically analysed to achieve reasonable understanding of the current scenario.

Awareness of Elder Abuse—The Tip of an Iceberg

The professionals who participated in the study are aware of elder abuse. However, their understanding of it has been gained through their own initiative and not through any formal training. All but one were aware of elder abuse as a phenomenon and its prevalence in the Indian context. The exception was a medical practitioner who felt that elder abuse does not happen in the Indian context and does not befit Indian culture.

Through their narratives it was realized that awareness of the term and phenomenon 'elder abuse' does not mean that there is clarity on what it constitutes. Awareness of the types of abuse that are often articulated in the academic literature were lacking amongst the professionals. None of them could articulate neglect and verbal abuse as abuse. Somehow, 'physical abuse' was often given the status of abuse and other categorizations were often seen as a normalized process of family dynamics: 'Everyone experiences it, everyone cannot be happy in a family all the time.' 'All of us know that inter-generational conflict does arise in a joint family.'

There were differences in the way the professionals perceived abuse. Whilst the counsellors and social workers could articulate the types of abuse (physical, emotional, verbal, sexual and self-neglect), it was interesting to note that the doctors could not. This was essentially because in the medical practitioners' understanding, 'abuse' was equated only with physical harm and injury.

The researchers could see that the counsellors and social workers could draw the implications of the stress associated with emotional abuse and neglect. There was recognition by counsellors of dimensions of abuse other than the physical kind. A great deal happens around the household space. The elderly are able to cope well with compromising on spaces like the living room, on television and bed arrangements, but they feel distressed when it comes to the kitchen. Two scenarios are common: the elderly are expected to work more in the kitchen, or they are denied use of the kitchen space. The counsellors and social workers have also narrated instances that have caused emotional distress to the elderly when they were not allowed to talk to the grandchildren or even to share with them any meal that they had prepared. Such restriction of the grandchildren's interaction with the grandparents is becoming common in many households, and causes huge emotional distress for the elderly. It is an intentional attempt by adults to hurt the elderly or use this as an emotional pressure to attain what they want.

The counsellors and social workers also indicated the distress associated with food and kitchen. Older women have articulated the distress when they are forbidden to cook food for the family. The worst stress described by practitioners was caused by having two separate kitchens in one household. Food being denied even on those days when the elderly have not prepared a meal because of their ill-health has also been described by practitioners. However, there were also documented experiences of older women being made to work in the household in excess of their physical capacity. Such instances have arisen because of lack of compatibility between two generations. Practitioners also concluded that unfortunately this will not be perceived as abuse, even if it obviously has physical and mental health concerns associated with it. Elderly people, because of their age-related vulnerabilities, withdraw from their routine eating habits, which results in a downward spiral for their health.

> I have seen them crying in my office, that they are not allowed in the kitchen and they feel so sad that they are unable to cook anything in the kitchen. (Reported by counsellor.)

> He was not brought to me for counselling, but for an advice as he is so resistant to any feedback and instructions by the family members and that he stopped eating and talking to people. After spending substantial time I got to understand that he was made to feel useless and unimportant. The food will be served and kept, but none checks with him, if he has eaten or not eaten, the bed linen and his clothes are not washed and changed for weeks. He does not want to make it as a complaint, but the whole thought of being a burden in the family made him to withdraw from food, which resulted in ill health. (Reported by social worker.)

> The older women sometimes are caught between their spouse and adult children. One of my clients reported that she gets verbally abused for causing disturbance if she gets up early to make a cup of tea for her husband who is so accustomed to have a cup of tea early in the morning. (Reported by counsellor.)

The medical practitioners in their interviews reported that they receive elderly clients who come to their clinics with complaints like malnutrition, urinary incontinence (UI), vaginal bleeding, and other age-related health problems such as high blood pressure and diabetes. There is strong interconnectedness between these physical problems and abuse. But they mentioned that the correlations are not established in their practice. They also could not draw the linkage between withdrawals from medication/overdosage of medication and abuse. Malnutrition can be associated with self-neglect, vaginal bleeding can be related to sexual abuse. Amongst the various conditions that have been repeatedly identified in the literature as being comorbid with UI, the most notable psychiatric condition continues to be depression. However the physicians articulated that they do not explore the reasons and they focus only on the presenting conditions. Unfortunately, none of the practitioners has paid attention to their patients' mental health problems, memory and emotions.

Psychiatry-based theories for the development of the co-morbidity in question seem reasonable, and a more innovative approach suggests that the neurochemical dysfunctions associated with depression may, in some way, also contribute to issues related to micturition (Steers and Lee 2001). Elderly people with signs of abuse may or may not have experienced incidents evidently attributable to abuse; those who have not been subjected to abuse could have symptoms that mimic abuse, and the signs and symptoms of many illnesses of late life can be erroneously ascribed to it. The professionals need to make these associations, so that it becomes the first step towards intervention. They need to evolve skills with which to fathom what is not spoken by the elderly. The doctors felt that it is challenging to probe on these lines as the elderly are often accompanied either by their children or in-laws. They also feel that it is not part of their expected role.

Thus, this section of the chapter can be concluded by saying that the understanding of elder abuse is just the tip of an iceberg, and the knowledge possessed by medical practitioners is generic rather than nuanced. None of the practitioners has described inputs or formal training on elder abuse in their curriculum. The understanding of each practitioner is dependent on the exposure they have had and the clients they have seen. Whilst the social workers and counsellors have been able to observe, identify and intervene to some extent, the medical practitioners have not been able to identify, and they have their own reasons. The doctors possess understanding and information, but have not necessarily translated it into practice as they lack time and skills to identify and intervene.

Interventions—Blazing a Trail

The practitioners, without any reservations, agreed on the need for interventions for elder abuse in health-care spaces. The realization that nothing much is happening is agreed upon unanimously. All of them in their practice have intervened on behalf of seniors, but only two mentioned that they had engaged with elderly victims who were suffering abuse. Though they have not received any training or specific inputs, they have intervened with their own generic understanding and used guidelines of family therapy. However, they were not sure if the problems were resolved as there was no follow-up afterwards. To start with, training in this field is practically non-existent.

> There is no content on elder abuse in the post graduation curriculum of counselling psychology. The content on elderly is around care-givers and dementia'. (According to a counsellor who has more than a decade of practice.)

> I also have not seen any training that happens in the field of geriatric counselling' (Counsellor.)

> I visit colleges to teach undergraduate psychology students. I do not see any inputs on elderly and concerns around elderly. (practitioner, who also is a guest faculty in teaching programs.)

The social workers said that not much emphasis is given on this theme as part of their training and capacity building. But they always handle the elderly issues with utmost sensitivity. A social worker stated that they do not see cases related to elderly people in isolation, but they have a humane approach to all the clients. She also added that, being located in an overcrowded public hospital, there is less time to address the psycho-social concerns. There were a lot of variations in the responses of the social workers. One of them stated that they 'have never used the lens of elder abuse'. On the other hand, another social worker stated that 'we do family sessions when we identify ongoing abuse'. According to one of them, the interventions are at times voluntary and not expected to be done as part of their job. In most cases at the workspace, it was also noted that the interventions done by the professionals are based on their individual choice and willingness to help. At least two social workers indicated that they have to establish a strong rapport with the elderly and only then do the elderly share their experiences related to abuse. Otherwise the elderly will never open up: 'Winning their trust is crucial for the elderly to open up about abuse'.

It can also be inferred that there is a lack of protocols regarding identification of elder abuse, intervention and a mandated referral mechanism. Also, for an effective referral mechanism, there should be services in place. There is paucity of such facilities for both poor and rich. The practitioners, who have engaged actively with victims of elder abuse, narrated that they do have no option but to advising them to adjust and cope with the dynamics within the family. They could find no alternative space to the abusive home that could provide them with shelter and food.

To conclude, it is encouraging to see that the public hospitals do run geriatric clinics on a weekly basis with a multi-disciplinary team. However, it is of concern that they have still not evolved protocol for elder abuse intervention. Despite the lack of system, protocol and training, it was very positive to see the willingness and motivation of practitioners to engage on this issue. The geriatric clinic and the multi-disciplinary team constitute a good beginning, but this space should not restrict itself to the treatment of clinical ailments, but rather be a comprehensive space for elderly care. The counsellors also felt that if the health-care and social-services sectors are not well equipped to identify and deal with the problem, elder abuse will continue to be under-diagnosed and overlooked. There is an urgent need to evolve protocols for assessment, training and sensitization of professionals, creating shelters and day-care centres for seniors.

Barriers and Challenges

The practitioners described many concerns about the barriers and challenges that influence screening, detection and intervention in the context of elder abuse. The reasons for these range from macro to micro systems.

The micro unit is the elderly themselves, who are not keen to share their experiences of abuse. There is shame and stigma associated with it. In addition, it compromises the family's name and reputation in society at large. Financial dependency on the adult children emerged as a major barrier to seeking help. The elderly also do not want to talk against the only support system they have. Regarding the financial concerns of the elderly, one of the counsellors stated that, 'they cannot afford the therapy, because they have to borrow money from children. … I had a client who felt good after sharing her feelings; however, she was not comfortable paying Rs. 1000 as, in her opinion it was a chit chat'. It also is partly because of their belief and value system in not 'wasting' money on themselves.

Furthermore, there is a lack of awareness among elderly people about the spaces that are available for them in which to seek support in times of crisis. Age was also cited as one major reason for not seeking help, as elderly people never feel that the young professionals can understand their problems.

Apart from these micro factors, lack of facilities, trained personnel and challenges related to accessibility were also reported. Rapport is crucial for the elderly to open up and unfortunately, because of the workload, the medical practitioners cannot spend much time on individuals. One practitioner said that the old concept of a family doctor worked well for the elderly because they know the family dynamics. 'Today in the super specialized mode, we have lost out on the human connect that existed before between practitioners and patients.'

Accessibility of services as a barrier was cited by almost everyone. Most elderly people are housebound and thus cannot move out to access services. Few studies have cited the highest prevalence of abuse amongst housebound dependent seniors. Even if they are mobile, they cannot navigate the complex health system on their own and thus the relatives must accompany them. This becomes a huge hindrance. 'We still do not have aged-friendly services.' 'We do not have active outreach, no volunteers visit housebound on a regular basis.' Other than this, older women experience certain emotions that act as a barrier to reaching out for support.

Most of the findings are similar to those reported by Beaulaurier et al. (2005), who described six key factors for older abused women that were barriers to them seeking help: self-blame, powerlessness, hopelessness, the need to protect the family, and the need to keep the abuse a secret from others. Also, their financial dependency on the family is cited as a major reason. Older battered women also explore the financial and practical realities of leaving an abuser, such as where they will live and how they will pay for insurance and medications. On the one hand they have listed the barriers to reaching out from the elderly people's point of view, but they have also shared the challenges the systems pose. Care of the elderly has never been a primary concern. One of the social workers from a specialized hospital said that they do not have a special unit for geriatric care, but they do have one for paediatric care. The government facilities do have a geriatric clinic which functions twice a week. But it is either on the first floor or on the second floor and is thus inaccessible for elderly people who have mobility-related issues. The counsellors also said that their set-ups are not very senior-friendly. Some stated that this is not just a problem related to structures, but also to the nature of the training they have received. A component on elder abuse has not been incorporated in the training given to the counsellors. They have not received any formal short-term or long-term training on this. Due to lack of assessment skills and lack of understanding on the depth and manifestation of the problem, it is impossible to effectively engage and intervene on issues around elder abuse. As of now, as the medical practitioners said, they do not have a peer-review mechanism to discuss with a multi-disciplinary team. Within the health-care context, elder abuse can be identified—or ignored altogether. This is the reality in the Indian context. Even if identification happens, there is no appropriate referral network or mechanisms to address the grievances. In cases where the elderly suffer violence in the familial context, they still remain within the family because we have not created formal spaces that can replace them. Thus, the practitioners often tend to ignore the larger issue and play a minimal role. WHO (2008) documented that primary health-care (PHC) workers are in an ideal position to recognize, manage and help prevent elder abuse and neglect. However, in reality we see that most of them do not detect it, as this is not part of their formal training. PHCs, legal and social services are ill equipped to identify and deal with this problem. In rural contexts, elderly people do not even reach the PHC facility because of the distance. Although awareness of the problem has increased in the past few years, elder abuse continues to be under-diagnosed and overlooked. The other significant gap, as shared by the practitioners, is the standardized tool to assess elder abuse. WHO (2008) has documented that elder abuse takes place within a context and, without a comprehensive assessment of the bio/psycho-social context of elder abuse, any screening or assessment instrument has significant limitations (Fulmer et al. 2004). An effective tool would be one that uses an inter-disciplinary approach and participatory research from the start of its development. The more disciplines and older people involved in the design and refinement of a tool, the better it will be accepted by both professionals and patients. It will also improve the inter-professional practice on referrals and interventions.

Recommendations: The Way Forward

Health-care practitioners, counsellors, social workers and lawyers are in a uniquely crucial position for the identification and treatment of abuse. Interventions by social workers, using the following two models, can stop or prevent the abuse: the Control Model, which talks about legal intervention; and the Support Model, which aims at therapeutic intervention or counselling (Yechezkel and Liat Ayalon 2013). Under the Control Model, abuse is looked at as a crime and legal means are employed to keep the perpetrator in check. The Support Model, on the other hand, takes into account the personal situation of the aggressor and relies on therapeutic interventions largely through dialogue, mediation, group therapy and support services such as domestic help and day-care centres (Gelles cited in Yechezkel and Liat Ayalon 2013). Both models are applicable for all age groups. According to Preston-Shoot and Wigley (2002), social workers most commonly adopt a supportive and therapeutic approach. In the absence of the Control Model, activating the Support Model will help the elderly.

Supportive interventions can be planned using an ecosystems framework in order to bring changes at micro, mezzo and macro levels. First and foremost, it is crucial to bring the discourse around abuse out of the family domain and into a public domain. There should be systematic surveys and studies to capture the typology, patterns, intensity and magnitude of the problem. The research should help us to strategize interventions that address elder abuse. Research by professionals should help in creating protocols for diverse practitioners to assess, intervene and make appropriate referrals. For referrals there is a need to create many forms of services such as day-care, transit-care and specialized care facilities.

In-service training on elder abuse and capacity building of diverse stakeholders in the health system and among community-based health workers will help in addressing the issue. More and more sensitization campaigns at the larger level will help to minimize the internalized stigma among elderly people, and also create vigilance on abuse in society at large.

Participant's practice-derived recommendations were similar to these. They cited the need for training and capacity building. The counsellors clearly indicated that it is important to include 'geriatric care and elder abuse' in the curriculum. As many practitioners still engage with academia as resource persons, they are aware that this is not inbuilt in the curriculum yet. The social workers also felt the need for training; however, one mentioned that they go through in-house training on handling sensitive issues. Professional training for practitioners is important as it results in: increased levels of knowledge, confidence in identification, and skills of assessment and intervention. However, little change will occur without situating abuse in the broader socio-cultural context, and without structural changes that allow for continuing education, supervision and support. An in-built referral mechanism assessing victims according to their need should be in place. The skills of networking and referral are important skills for practitioners to acquire. The second element is conducting research to evolve a nuanced understanding of elder abuse and its ramifications. Even today there is no clarity on what constitutes elder abuse in a familial context or on the measures that are in place to combat this problem. Research has indicated that a multi-pronged approach is needed, but further participatory studies need to be conducted to understand and delineate this. We have to evolve a context-specific, comprehensive assessment tool for the health-care workers. It is surprising to see that, despite so much data on prevalence, we have not built mechanisms to address elder abuse. A referral networking system has also to be established. Lessons have to be learnt from existing good practices in other countries. Outreach initiatives and home-visits should also be made mandatory, as there are many housebound elderly people in the Indian context.

Conclusion

Elder abuse in any form is a violation of older people's human rights. Studies and surveys from India definitely indicate the alarming increase of the problems here. This chapter concludes by saying that we have not evolved appropriate mechanisms to address elder abuse. The positive dimension is that there is openness and willingness on the part of health-care practitioners, social workers and counsellors to play an active role in the elder abuse intervention process. Thus, it is important to engage this professional space for elder abuse interventions. Professionals should be trained and their capacities should be optimally utilized to address the concerns around elder abuse. There are health and mental health concerns around elder abuse, and thus positioning the elder abuse intervention within these non-threatening spaces befits the cause. There are earlier examples of addressing violence in familial contexts through these kind of spaces. Learning lessons from such models, we should move forward to an senior-friendly environment.

References

Bagshaw. 2011. Equal or different? Respecting older women: Responding to older women's experiences of family violence, Sexual Assault and elder abuse. *Domestic violence resource centre conference Melbourne*, 6 April 2011.

Bagshaw, Dale, Sarah Wendt, Lana Zannettino, and Valerie Adams. 2013. Financial abuse of older people by family members: Views and experiences of older Australians and their family members. *Australian Social Work* 66 (1): 86–103.

Beaulaurier, Richard L., Laura R. Seff, Frederick L. Newman, and Burton Dunlop. 2005. Internal barriers to help seeking for middle-aged and older women who experience intimate partner violence. *Journal of Elder Abuse & Neglect* 17 (3): 53–74.

Brandl, Bonnie, and Tess Meuer. 2000. Domestic abuse in later life. *Elder LJ* 8: 297.

Breaux, John B., and Orrin G. Hatch. 2003. Confronting elder abuse, neglect, and exploitation: The need for elder justice legislation. *Elder LJ* 11: 207.

Buchbinder, Eli, and Tova Winterstein. 2003. 'Like a wounded bird': Older battered women's life experiences with intimate violence. *Journal of Elder Abuse & Neglect* 15 (2): 23–44.

Carney, Maria T., Fay S. Kahan, and Barbara E. Paris. 2003. Elder abuse: Is every bruise a sign of abuse? *Mount Sinai Journal of Medicine* 70 (2): 69–74.

DESA, UN. United Nations Department of Economic and Social Affairs/Population Division. 2015. World Population Ageing 2015 (ST/ESA/SER.A/390).

Fulmer, Terry, Lisa Guadagno, and Marie Therese Connolly. 2004. Progress in elder abuse screening and assessment instruments. *Journal of the American Geriatrics Society* 52 (2): 297–304.

India, HelpAge. 2013. *Elder abuse in India* (2014).

James, Marianne Pinkerton, and Adam Graycar. 2000. *Preventing crime against older Australians*. Canberra, ACT: Australian Institute of Criminology.

Killick, Campbell, and Brian J. Taylor. 2012. Judgements of social care professionals on elder abuse referrals: A factorial survey. *British Journal of Social Work* 42 (5): 814–832.

Krug, Etienne G., James A. Mercy, Linda L. Dahlberg, and Anthony B. Zwi. 2002. The world report on violence and health. *The Lancet* 360 (9339): 1083–1088.

Lachs, Mark S. 2004. Screening for family violence: What's an evidence-based doctor to do? *Annals of Internal Medicine* 140 (5): 399–400.

Nahmiash, Daphne, and Myrna Reis. 2001. Most successful intervention strategies for abused older adults. *Journal of Elder Abuse & Neglect* 12 (3–4): 53–70.

Perel-Levin, Silvia, and World Health Organization. 2008. Discussing screening for elder abuse at primary health care level by Silvia Perel-Levin.

Preston-Shoot, Michael., and Veronica Wigley. 2002. Closing the circle: Social workers' responses to multiiagency procedures on older age abuse. *British Journal of Social Work* 32 (3): 299–320.

Raju, S. Siva. 2002. Meeting the needs of the older poor and excluded in India. *Situation and Voices* 93.

Schiamberg, Lawrence B., and Daphna Gans. 2000. Elder abuse by adult children: An applied ecological framework for understanding contextual risk factors and the intergenerational character of quality of life. *The International Journal of Aging and Human Development* 50 (4): 329–359.

Schmeidel, Amy N., Jeanette M. Daly, Marcy E. Rosenbaum, Gretchen A. Schmuch, and Gerald J. Jogerst. 2012. Health care professionals' perspectives on barriers to elder abuse detection and reporting in primary care settings. *Journal of Elder Abuse & Neglect* 24 (1): 17–36.

Skirbekk, Vegard, and K.S. James. 2014. Abuse against elderly in India—The role of education. *BMC Public Health* 14 (1): 336.

Steers, William D., and Kyu-Sung Lee. 2001. Depression and incontinence. *World Journal of Urology* 19 (5): 351–357.

Teaster, Pamela B., and Tenzin Wangmo. 2009. *Elder abuse detection and intervention: A collaborative approach by B. Brandl, CB Dyer, CJ Heisler, MJ Otto, LA Stiegel, and RW Thomas (2007)*, 83–85. New York: Springer.

Thomas, Cynthia. 2000. The first national study of elder abuse and neglect: Contrast with results from other studies. *Journal of Elder Abuse & Neglect* 12 (1): 1–14.

United Nations. 2015. *World population prospects: The 2015 revision*. http://esa.un.org/unpd/wpp/.

van Gerwen, Maaike, François Schellevis, and Toine Lagro-Janssen. 2007. Comorbidities associated with urinary incontinence: A case–control study from the second Dutch National Survey of General Practice. *The Journal of the American Board of Family Medicine* 20 (6): 608–610.

Vinton, Linda. 2001. Violence against older women. *Sourcebook of Violence Against Women* 179–192.

Vinton, Linda. 1992. Battered women's shelters and older women: The Florida experience. *Journal of Family Violence* 7 (1): 63–72.

World Health Organization. 2002. Missing voices: Views of older persons on elder abuse. In *Missing voices: Views of older persons on elder abuse*, p. 23.

Yechezkel, R., and L. Ayalon. 2013. Attitudes towards Intimate Partner Abuse in Younger vs. Older Women. *Journal of Family Violence* 28 (4): 381–391.

Zwi, Anthony B., Etienne G. Krug, James A. Mercy, and Linda L. Dahlberg. 2002. World report on violence and health—Exploring Australian responses. *Australian and New Zealand Journal of Public Health* 26 (5): 405–408.

Chapter 14
Prevention and Management of Elder Abuse

Ruchika Kuba

Introduction

Elder abuse is defined by the World Health Organization (WHO) as a 'single or repeated act, or lack of appropriate action, occurring within any relationship where there is an expectation of trust which causes harm or distress to an older person' (World Health Organization, Geneva 2008).

Abuse is a global phenomenon which occurs irrespective of religion, gender, culture and economic status of the society. Ageing may trigger an additional risk of abuse due to the increased dependence on others, social isolation and frailty that accompany it (World Health Organization, Geneva 2008). Abuse, especially of older persons, may not come to light because the symptoms can be confused with changes brought about by ageing. Another reason why abuse or neglect may go unnoticed is that sometimes elderly people choose not to report it because of family pride or self-blame. Moreover, older men and women come from generations that avoided discussing private issues. As a result, elder abuse continues to be a taboo, mostly underestimated and ignored by societies across the world (World Health Organization, Geneva 2008).

Prevalence rates/estimates do not exist in all countries and have so far generally been restricted to a few developed nations. Where there are prevalence studies on elder abuse, rates range between 1 and 35% (Pillemer and Finkelhor 1988; Ruiz Sanmartín et al. 2001; Yan and Tang 2001) depending on definitions and survey and sample methods. These figures, however, may represent only the tip of the iceberg, and some experts believe that elder abuse is under-reported by as much as 80%. Estimates of the number of elder abuse cases reported by other studies range from 1 in 15 cases to 1 in 6 cases. The rates in all these studies are relatively low, and more proactive studies are required to find the real incidence. These low rates

R. Kuba (✉)
School of Health Sciences, Indira Gandhi National Open University, New Delhi, India
e-mail: ruchikakuba@yahoo.com

Mala Kapur Shankardass and S. Irudaya Rajan (eds.), *Abuse and Neglect of the Elderly in India*, https://doi.org/10.1007/978-981-10-6116-5_14

may be due to the isolation of older people, the lack of uniform reporting laws and the general resistance of people—including professionals—to report suspected cases of elder abuse and neglect. In developing countries, although there is no systematic collection of statistics or prevalence studies, crime and social welfare records, journalistic reports and small-scale studies provide evidence that abuse, neglect and financial exploitation of older people appear to be widely prevalent (World Health Organization, Geneva 2008).

Except for census enumeration, the government of India lacks collection of information on factors related to elderly people such as prevalence and incidents of elderly abuse and neglect. HelpAge India in its report (Soneja 2011) on elder abuse in India presented statistical data on the distribution of abuse in major cities in India. It discussed the types of abuse in the different cities, the responses of the elderly to abuse and their perception of the laws, policies and role of police and lawyers.

The advocacy work done by the International Network for Prevention of Elder Abuse (INPEA) and the emphasis given to elder abuse prevention by the WHO have contributed significantly to raising worldwide awareness of elder abuse. Academic institutions around the world have also substantially contributed to enhancing understanding and raising awareness. and have developed methodological tools to study the problem. However, much is still to be done (World Health Organization, Geneva 2002b).

Causes and Risk Factors

Before we talk about prevention and management, it is important that we discuss the causes of abuse. There are not many research studies done on this. Also the causes will vary according to social strata. For example, issues pertaining to property are generally seen as the major cause for elder abuse in developing countries. In the developed countries the causes may relate more to the isolation of the older generation. Older adults frequently find themselves mistreated in various ways by people they trust, with significant, lasting consequences (World Health Organization, Geneva 2015). The following list gives some of the causes in the Indian context. It is important to focus here on the causes which result in a multiplier effect on the actual/reported cases of elder abuse.

1. *Changing family structures and the position of the elderly*
   Elderly people are slowly losing their position of authority in the family. Their views are usually not asked and their interference is not welcome and often rebuked. The moving-out of the younger generation and development of nuclear families often leave the elderly isolated in the place of family roots, and hence they suffer neglect. Erosion of family bonds and a history of abuse or violence in the family are risk factors which increase the chances of abuse.

2. *Deterioration of the physical and mental health of the elderly*
   With the increase in life expectancy and longevity, the number of dependant elderly is increasing. The physical and mental ailments make seniors more prone to abuse as their ability to defend themselves is reduced.
3. *Care-giver's stress*
   The increasing burden of caring for the elderly can result in physical and mental strain. These can at times be expressed in the form of violence or neglect, thus giving rise to abuse. The stress of care-giving can be increased by other stresses on the care-giver, deriving from time constraint, work or family, for example.
4. *Psychopathology of the abuser*
   Various psychological factors such as personality disorders, mental disorders, alcohol or drug addiction can increase the chances of elder abuse, as can low moral values.
5. *Attitude of the elderly towards abuse*
   The elderly may not report abuse, which in turn increases the abuse.
6. *Other miscellaneous factors*

   There could include:

- non-availability of framework for out-of-court settlement
- lack of professional counselling
- non-enforcement of serious penalties
- lack of monitoring and reporting by the local police
- absence of police departments specifically for elderly people similar to the ones available for women, children, economic offences etc.

Effect of Abuse

The WHO, in its 2002 policy framework (World Health Organization, Geneva 2002b), mentioned that elder abuse is a violation of human rights and a significant cause of injury, illness, lost productivity, isolation and despair.

The global response to elderly abuse and neglect in the WHO 2008 Report (World Health Organization, Geneva 2008) observes that elder abuse has serious consequences for the health and well-being of older people and can be of various forms: physical, verbal, psychological or emotional, sexual and financial. It can also simply reflect intentional or unintentional neglect. Abuse and neglect are culturally defined phenomena that reflect distinctions between value standards and unacceptable personal behaviour.

Studies have found that older people perceive abuse under three broad areas: neglect (isolation, abandonment and social exclusion), violation (of human, legal and medical rights) and deprivation (of choices, decisions, status, finances and respect) (World Health Organization, Geneva 2002c).

Signs and Symptoms of Abuse

The signs and symptoms of abuse depend upon the type of abuse. Some are listed below.

1. *Physical*
   - signs of injury which cannot be explained, such as cuts, bruises, fractures
   - injuries in different stages of healing
   - overdosage of drugs or missing dosages
   - delay between the occurrence of injury and reporting
2. *Emotional*
   - signs of insecurity and psychological disturbance, like thumb sucking, rocking
   - abusive behaviour of the abuser witnessed by the doctor
   - depression, confusion, fearfulness
   - sudden change in behaviour or alertness
3. *Sexual*
   - bruises around genitals
   - unexplained vaginal or anal bleeding
   - sexually transmitted infection
4. *Neglect*
   - shabby appearance and dirty clothes
   - unexplained weight loss and or dehydration.

Need for Prevention and Management

Prevention and management of elder abuse are of paramount importance, since abuse has an impact on all aspects of a senior's life, ranging from physical, mental and psychological areas to social, sexual and financial ones. However it is a complex issue and requires a multi-faceted approach. Older persons can themselves be involved in designing programmes to spread awareness and prevent abuse. Besides increasing awareness, some proactive steps need to be taken. Commitment is required at all levels. At the national level, appropriate policy, laws and programmes must be successfully implemented. Action is also needed at the institutional and community level. It is essential for society to become aware of elderly people's rights and to recognize the existence of elder abuse or neglect. Involvement of the elderly in prevention and management of abuse is an important aspect. Seniors can be empowered with the knowledge of their social, legal and financial rights. It is equally important to focus on the research component associated with the abuse of older persons. Studies are required to look into the causes, risk factors, and the roles of culture and care-givers' stress, among many other factors.

Global Responses

The Toronto Declaration on Global Prevention of Elder Abuse 2002 (World Health Organization, Geneva 2002d) is a Call for Action aimed at the prevention of elder abuse. This declaration was devised at an expert meeting, sponsored by the Ontario Government in Toronto on 17 November 2002. According to this declaration, elder abuse is a universal problem, prevalent in both developed and developing countries. However, it is more common in high-risk groups such as the very old, women, the poor and those with limited functional capacities. The Declaration advocated primary health-care (PHC) workers playing an active role in the prevention and management of abuse. To prevent abuse there is a need to increase awareness and education about the issue among the general public, media and the professional sector. There is also a need for all countries to provide health and social services, legal protection, police referral, etc., and to appropriately respond and eventually prevent the problem.

The United Nations International Plan of Action adopted by all countries in Madrid (April 2002) clearly recognize is the importance of elder abuse and puts it within the framework of Universal Human Rights. It advocates that everyone should be involved in preventing elder abuse.

The response of different countries (World Health Organization, Geneva 2002) around the world to developing legislation and other policy initiatives is at different stages. Responses at national level use Blumer's model (Blumer 1971) of social problems to describe progress.

1. Emergence of a problem.
2. Legitimization of the problem.
3. Mobilization of action.
4. Formulation of an official plan.
5. Implementation of the plan.

The developed countries have a better established system for reporting and treating elder abuse. In countries like the USA and the United Kingdom, there is a national policy in place. On the other hand, countries like Australia and Canada have no national policy, although they have set up systems at the state level and in some provinces. New Zealand has a National Elder Abuse and Neglect Advisory Council, formed in the 1990s, for providing national perspective on strategies for care and protection of older persons. Other European countries (such as France, Germany,

Italy and Poland) have activities related to elder abuse and prevention through individual researchers and local programmes. In South America, countries like Argentina, Brazil and Chile have moved on to legitimization and action. However, in countries like Cuba, Peru, Uruguay and Venezuela, awareness of the problem is still emerging. Support services, including a telephone helpline, are available in Argentina. In the Asian countries, researchers have helped to increase awareness of elder abuse in China, Hong Kong, India, Japan etc. The government of South Africa is considering a national strategy on elder abuse.

INPEA, formed in 1997 has representatives from all six continents. It works towards reducing the cases of elder abuse and aims to increase public awareness, education, training and research. It also campaigns on behalf of the abused or neglected people. Its network achieves these objectives by organizing workshops, setting up professional meetings and training sessions, maintaining a website and producing a newsletter.

Role of Professionals and Support Staff

Most elderly people suffer from one or more ailments and come in contact with medical professionals or other health-care personnel. This is an excellent opportunity to detect cases of elder abuse. The elderly place their trust in doctors and are more likely to open up and tell about their abuse. It should be expected that doctors at the PHC level or general practitioners will be able to detect and manage the abuse appropriately. They can address the medical issues and also provide counselling. However, it has been found that they generally hesitate to intervene and prefer to refer the cases to a social worker. Health professionals rarely report cases of abuse (Lachs et al. 1997; Clark-Daniels et al. 1990; Barer 1997).

In a survey (Jones et al. 1997) of accident and emergency ward physicians, it was found that only 31% were aware of a written protocol for elder abuse and most were unfamiliar with the reporting mandates for elder abuse in their state. Surprisingly, only 2% of all elder abuse reports are generated by physicians, despite state-wide mandates in most states (Rosenblatt et al. 1996).

Such inert behaviour could be due to lack of knowledge and training in this area. Another reason could be lack of time due to the heavy caseloads doctors are required to process. Studies have also shown that the various health-care professionals (including doctors, nurses and other staff) feel that they are not qualified or competent to assess the elderly for signs of abuse or neglect (Stotkowski 2008; Wagenaar et al. 2009). In one survey, two-thirds of residency programmes did not place a major emphasis on training about elder abuse (Wagenaar et al. 2009).

Difficulties in the assessment of elder abuse have been studied (Lachs et al. 1997) and include factors such as: (a) no sensitization or training; (b) lack of communication and coordination among the different medical personnel; (c) no set protocols for standard interventions; (d) absence of specific definitions of abuse and its associated terminologies; (e) lack of any social support existing for the care-givers; and (f) no awareness or information available regarding the presence of institutional resources.

However, as mandatory reporters of abuse and neglect, health-care providers have both an ethical and legal responsibility to advocate for victims of abuse by screening, identifying and reporting cases of abuse (Shugarman et al. 2003; Bomba 2006).

It is therefore important that the sensitization of the PHC professionals be increased in all countries around the world. Equally important is the laying-down of legislations and guidelines regarding the management of elder abuse. In a few countries (such as Costa Rica, Kenya and Singapore) legislation does exist; however, it does not cover issues of elder abuse adequately. Brazil has mandatory reporting, but concerns were raised on behalf of PHC professionals who feared for their own safety (World Health Organization, Geneva 2008).

The second important step is to develop PHC professionals' and social workers' capacities to deal with elder abuse. Various initiatives have been suggested (Lachs et al. 1997): a few of them are outlined below

- Sensitization of governments regarding issues related to elder abuse. This will enable them to engage PHC professionals, including physicians.
- Medical professionals should be trained regarding detection and management of the abuse. The specific roles of different professionals (such as doctors, nurses and para-professionals) in a hospital need to be outlined.
- Medical professionals must have guidelines regarding appropriate referral of such cases to social workers or other organizations such as non-governmental organizations (NGOs) where proper management of the cases can be done.
- Awareness levels in the community should be increased to help prevent the problem, detect cases, increase reporting and provide support mechanisms.
- Elder people should be informed about their rights, especially in relation to abuse, neglect and exploitation.
- The person abusing the elder, needs also be included in the management plan so that the recurrences do not occur and a permanent solution can be arrived at.

Education of physicians, especially at primary-care level, can increase their competencies in identifying the signs and symptoms of elder abuse.

Some studies based on survey results (Wagenaar et al. 2010) have shown that doctors who have been oriented or trained in the area of elder abuse during their residency, or who have participated in continuing medical education activities, then considered themselves able to detect elder abuse. It is to be wished that such sensitization and awareness regarding elder abuse should be made part of the educational activities in primary care disciplines. The subject should also find a place in general articles, audio tapes and primary lectures (Taylor et al. 2006; U.S. Preventative Services Task Force 2004).

With the advances in technology and the availability of online resources, programmes, massive open online courses (MOOCs) and continuing medical education programmes could also be prepared for physicians (Hill 2005). Jones et al. (1997) found that 92% of physicians surveyed did not believe that their states had sufficient resources to meet the needs of elderly victims; this belief could contribute to reporting apathy.

The physicians have given suggestions regarding the areas to be included in the training programmes. Some of these areas include the law and clarity on reporting procedures (Dumortier and de Freminville 2014). They also desired a requirement to investigate the condition of the patient for possibility of abuse (Lachs and Pillemer 1995; Jones 1990). Abuse should be suspected if a doctor or other health-care worker notices any of the following signs:

- delay in seeking medical attention after an injury or an illness
- history given by the patient is different to that provided by his care-giver or family member
- no definitive explanation of the injury or ill-health by either the patient of the family member or care-giver
- indifferent attitude of the family member or care-giver towards the injuries or the advice given by the doctor
- no care-giver or family member accompanying a functionally impaired elderly person
- laboratory findings inconsistent with the history provided
- worsening of an elderly person's chronic condition despite diagnosis, provision of a definitive management plan and resources available with the family member/care-giver.

Guidelines need to be drawn up for the doctor or health-care provider to use while examining the elderly person suspected of having been abused (Ansen and Breckman 1988). These could include the following points:

- If the elderly person is physically and mentally able to recall and respond, he should be interviewed alone and not in the presence of the family or care-giver. He should be directly asked about the possibilities of physical violence, restraints and neglect.
- The suspected or implicated abuser should also be interviewed alone to explore the possibilities of abuse.
- The patient's formal and informal social networks should be explored.
- A comprehensive assessment of the elderly person should be carried out. This would include the medical, functional and cognitive components.

Strategies proposed for involving the health professionals in the prevention and management of abuse in India include:

- incorporation of awareness about elder abuse in the medical and nursing curriculum at undergraduate level
- teaching the skills to recognize and manage (or refer appropriately) cases of abuse during medical training at undergraduate level
- conducting various in-service training programmes for sensitization and training of health professionals in diagnosing and managing cases of abuse, including emergency management of cases due to abuse

- developing online training and the embedding professional education resources about elder abuse into a range of training curricula
- developing treatment protocols and referral strategies for health-care institutions which include government dispensaries, PHCs hospitals and private nursing homes and hospitals.

Care-Givers' Support Groups

Various agencies also provide support for care-givers. These may offer one or more of the following services:

- providing care-givers as per the requirements
- support through online/social groups
- care-giving training
- medical support
- nursing support
- laboratory support
- nutritionist/dietician support
- physiotherapist or occupational therapist services
- home modification support.

The list of job responsibilities for in-home helpers or care-givers is very long. But they are paid very low wages. Hence, demand for them is much greater than the supply. Due to the low wages, applicants who come forward for care-giving jobs are often individuals with limited alternative employment options. These could include people with limited educational qualifications and training, immigrants, poor language skills etc. In recent years, advocates for elderly and disabled people, unions and workers' advocacy groups have started working together to call for improvements in workers' pay and working conditions (Nerenberg 2002a). Through such advocacy, it is hoped that more individuals will enrol and a better service be provided to seniors.

Training of In-Home Helpers

Elderly people with physical and cognitive impairments may require assistance for many activities of daily living. Family members provide this support in most of cases. However, in many countries now, assistance comes from professionally employed in-home helpers who take care of the basic needs of elderly people with dysfunctions. The terms used to describe helpers vary widely and include 'personal care assistants', 'attendants', 'chore workers', 'in-home support service workers', and 'homemakers' (Nerenberg 2002b). The in-home helpers or assistants may be paid by the person or family employing them or through private insurance agencies. At times they are paid through federal programmes. Agencies which employ them train the in-home helpers with basic skills to care for the elderly. They also orient them towards detecting abuse.

At times agencies may also hire other organizations to train their workers. As part of the training, the workers are given the basic information regarding the common diseases and disabilities affecting elderly people. They are given training on how to manage the daily needs of seniors with various physical and mental impairments, use of their assistive devices, managing their diet and medicine routines etc. They are also trained on how to help elderly people with financial management and shopping, and assist in dealing with banks etc. The most challenging part of the helper's training is handling the elderly during abusive, violent or abnormal behaviour. Common topics for the care-givers' programme offered by various organizations include topics like:

- managing activities of daily living
- reacting to emergency situations
- handling injuries and falls
- managing resistance to care
- handling catastrophic or extreme stressful situations
- mastering verbal and non-verbal communication and counselling skills
- managing a bedridden senior
- handling medication errors
- type of help available
- caring for a care-giver
- personal care
- nutrition
- medical aspects of caring
- emotional and intellectual well-being
- legal and financial affairs
- home safety
- when it is time to stop caring
- choosing a residential care facility
- how to work with other professionals and working in multi-disciplinary teams.

Hence, they can observe acts which may be abusive or indicate neglect. Since the in-home helpers are in close proximity to the elders, they win the trust of the senior and be able to discuss private issues with them and help them in abusive situations.

Not all the training programmes touch upon the issues of elder abuse, its identification, prevention or management, but there is an imperative need to do so. The training programmes should empower in-home trainers to:

- identify elder abuse in the families they work with
- protect themselves from being abused
- allay by personal example the general suspicion of helpers as potential abusers.

To develop a plan for elder abuse prevention and management, the following require to be identified at the planning stage:

- the needs felt by the participants
- the background of participants, to customize the training accordingly
- the elder-abuse reporting requirements of the state/country
- a policy which would need to be developed, based on the government's guidelines for managing abuse
- a procedure which would need to be put in place for handling abuse.

Some of the elder abuse-related areas that should be covered in training courses for in-home helpers might include:

- introduction to the topic of elder abuse (definitions, types, risk factors)
- assessment of abuse cases/recognizing and managing abuse (signs and symptoms of abuse)
- managing difficult behaviour which can give rise to abuse
- behaving in a professional manner
- details of local organizations that can provide assistance
- acquaintance with a standard plan outlined by their organization.

Care-givers' stress is one of the important risk factors for elder abuse. Research on elder abuse, carried out earlier, was based on the hypothesis that abuse or neglect result from either the stress of caring for a dependent elderly person or lack of relevant competencies and resources (Reingold 2006; Nerenberg 2008). Hence, spreading awareness, educating and training coupled with support programmes are essential components for prevention and management of elder abuse within a care-giving relationship (Wolfe 2003; Ross 2007). Various organizations and agencies thus work towards providing social and emotional support and guidance on how to handle difficult situations, violent behaviour and a potentially abusive environment. This reduces the stress among care-givers and brings down the chances of abuse. A large number of tools are now available by which the care-givers can assess their risks. Care-giver support groups play a multi-faceted role by providing social and emotional support, preparing the care-givers for their intended role, and training them to handle difficult situations (Nerenberg 2008).

Training can reduce the likelihood that workers, themselves, will engage in abusive or negligent conduct. Such components, when included in the training programmes, can help the workers by:

- providing skills to handle difficult situations, e.g. dealing with violent behaviour and learning to defuse potentially volatile situations with confidence
- empowering them to protect themselves from abusive, violent and inappropriate behaviour by their clients
- clarifying their doubts regarding the expectations of acceptable conduct towards clients
- providing them with the knowledge and skills to identify abuse and report it to the appropriate authorities for action

- helping them to handle the stress which is an inevitable part of care-giving
- sensitizing them to the penalties for abuse, and thus reducing the chances of abuse by care-givers.
- keeping their morale high and motivating them to continue the work sincerely even when they get no acknowledgment of the same.

Disturbing behaviours (Nerenberg 2002a) that are frequently encountered by in-home helpers include paranoia, aggression and combativeness. These behaviours are frequently a response to fear, frustration or the inability to communicate. Physical aggression often occurs while workers are providing personal care such as bathing. Tips for reducing disturbing behaviour are:

- maintain patience, avoid anger and reassure the elderly person
- be encouraging and avoid any arguments
- do not use force and avoid negative non-verbal communication
- look for signs of agitation or frustration
- assess situations for danger to self and the care-receiver
- politely refuse interference by other family members in your work
- for any problems outside your control, contact your supervisor.

The workers must practise in professional manner (Nerenberg 2002a). Practice principles give workers the tools they need to deal with difficult ethical situations. These help to keep the agencies and the workers protected from criticisms regarding their role in caring for the elderly. Some of the examples are given below:

- Do not take any money or gifts other than what is your salary.
- Be calm and patient with your client, even if he/she displays unruly behaviour.
- Stay away from, and do not encourage any, sexual advances irrespective of the patient's physical and mental condition.
- If the patient is uncooperative or does not listen to your repeated requests, report the matter to your supervisor.

The extent to which training interventions with care-givers may reduce the risk of abuse or neglect is still a matter of debate (Nerenberg 2008). Recent research has shown correlations between abuse and higher care burden (Cooper et al. 2009). However, further research is required to assess whether care-giving itself is a potential cause or risk factor for abuse, and the effectiveness of prevention and management of abuse with the help of care-givers (Pillemer et al. 2007).

General Awareness Programmes

Public education and awareness are important for preventing abuse and neglect. The objectives of an awareness programme are to:

- make the general public aware of elderly people's basic needs

- educate the public about the existence and signs of abuse and negligence, and the risk factors associated with them
- organize sensitizing programmes, workshops, seminars, training programmes etc. with the aim of changing attitudes and behaviour towards elderly people to prevent abuse
- help the public to identify signs and symptoms of abuse, where help can be obtained, and the responsibility of individuals in preventing and reporting elder abuse (World Health Organization, Geneva 2002a)
- promote the development of tolerance among the community towards the demands of elderly people.

The people targeted for these awareness and training programmes include doctors, nurses, other health personnel, police, social workers, legal functionaries, policy-makers, educators, researchers, elderly people and the general public. The programmes can be tailored keeping the target population in mind. They can be organized by different governmental organizations and NGOs, self-help and support groups, community centres, resident welfare associations etc.

Awareness programmes should be specific to region and culture. Abuse may be defined differently in different cultures. Non-reporting in different communities could have different reasons, such as seniors protecting their abusers, a sense of shame, stigma etc. Reports in an article on Native Americans in the USA state reveal that among the greatest obstacles to preventing abuse in is the reluctance of elders, their families, tribal leaders and others to admit that abuse is occurring or to seek help (Nerenberg 2004). The participants in a focus group study in South Africa stressed the importance of the media in raising public awareness (Keikelame 2000). Therefore, one must explore the possibilities to promote recognition of the existence of elder abuse and neglect, and the need to detect it, through various workshops for the general public which could be supported by governments and NGOs.

Awareness can also be spread through the social networking sites and media. This will help in changing the attitude of the community towards the elderly. These efforts will be constructive in preventing elder abuse (Srinivasan and Gupta 2015). Qualitative research with abused seniors found that often they are not aware of where to go for help. They also have little or no knowledge of their rights and entitlements. There is an awe of the authorities in general. They are also apprehensive that their issues and complaints may not be taken seriously by the various agencies (Pritchard 2000; Mowlam et al. 2007).

Social Support Services

The existing health and social networks are generally used to provide services to the elderly who suffer abuse or neglect. They mange the cases in a comprehensive manner and generally deal with the medical, legal, ethical, psycho-social, economic and environmental issues (World Health Organization, Geneva 2002d). The support groups often have people in their ambit who have had similar kinds of experiences and hence are able to understand and empathize with those experiencing or recovering from abuse (Nerenberg 2008). Support groups act as a source of affirmation, information and education, creating an empowering environment (Podnieks 1999).

Studies have found that greater social support is independently associated with a lower risk of self-reported mistreatment (Naughton et al. 2010). Those who are abused and have a lower degree of social support are believed to be at a higher level of psychological distress (Fulmer 2005; Dong and Simon 2008; Podnieks 2006). Hence, community-based support groups could be considered as effective interventions for both the abusers and the abused. (Podnieks 1999; Dong and Simon 2008; Hightower et al. 2006).

Telephone helplines to receive reports of mistreatment are often a feature of such systems (Action on Elder Abuse, London 1997; Yamada 1999). These helplines operate in a few European countries such as the United Kingdom, and in local communities in France, Germany and Japan. Programmes in San Francisco and Los Angeles have also experimented with small groups targeting perpetrators and using cognitive behavioural techniques to alter violent behaviour (Bonnie and Wallace 2003). Only the USA and a number of Canadian provinces have created a system solely for handling reports of adult mistreatment.

The concept of emergency shelters and support groups specifically aimed at the elderly has emerged recently. This provides a space where the abused seniors can discuss their experiences with each other. This helps them to build confidence, get psychological support, mange their anxiety levels and allay their fears and doubts. Such emergency shelters exist in Canada, Germany, Japan and the USA (World Health Organization, Geneva 2002d).

Programmes, for preventing abuse of the elderly in their homes, in which older people themselves play a leading role, include:

- developing and offering awareness and training programmes in which the elderly conduct the training programmes of the abused elders
- seniors being involved in providing company to their isolated peers
- setting up community programmes in which the elderly are encouraged to participate
- developing social networks for older people so that they can interact with, confide in and support each other
- establishing social support groups in which elderly people take part as members to help abused seniors to open up, share their experiences more easily, and accept suggestions and management
- creating self-help programmes in local areas to help seniors to become more productive for society, thus building up their self-esteem.

Technology and Assistive Devices

Assistive devices and use of technology to modify the home to suit the needs of the elderly help reduce care-givers' physical stress. They also improve seniors' quality of life and thus reduce the possibilities of abuse. Assistive devices include canes, walkers and wheelchairs. Home modification include ramps, broad electricity switches, hand rails, toilet seats at the correct height, proper lighting, bed alerts, fall detectors and door monitors. There are also smart technologies developed for monitoring elderly people and alerting care-givers.

Basic Rights of Elderly People

Older people's human rights must be guaranteed worldwide (World Health Organization, Geneva 2002a). Some ways in which this could be done are listed below:

- Modifications can be made in the already laid-down laws on domestic and intra-family violence to also cover the elderly.
- Exploitation, abuse and neglect of seniors can be also covered under the existing criminal and civil laws.
- New laws can be introduced by government exclusively to protect older people.

The elderly must be educated regarding abuse, its prevention and management. Some advisory suggestions are:

- Be aware of your rights.
- Plan your future, your daily plan, your contributions towards family and society, your finances and will.
- Be socially active and have your own group of friends and well-wishers.
- Seek out friends, authorities and agencies if you feel you are being abused.

The legal extensions and provisions will help give strength to the existing protective services and increase the number of support groups and other agencies engaged in the prevention and management of abuse, neglect and exploitation of the elderly. There is also a need to safeguard an individual's right to self-determination. This indicated their commitment to the doctrine of self-determination (Regehr and Antle 1997). The bringing in of resulted in removing a person's right to make his/her own decisions (Dyer et al. 2005). However, there is also an increasing trend globally to bring in legislative reforms to empower vulnerable adults to play an important role in decision making (Heath and Phair 2007; Donovan and Regehr 2010). Hence a legislative empowerment must be accompanied by the client's right to self-determination (O'Donnell 2012).

Legal interventions in cases of elder abuse may make use of both the criminal and civil justice system (Kalaga et al. 2007). The improvement in the legislative instruments and training programmes of the legal enforcement and prosecution officers has resulted in an increase in the number of cases of abuse being filed for prosecution (Nerenberg 2006).

The efficacy of criminal and civil justice interventions is also open to debate. A study by Filinson (1993) found that the recurrence rate for abuse was 24% for those who received assistance, support and advocacy in the use of the criminal justice system, compared to just 17% in a control group. In another study, the incidence of recurrence of abuse was unknown in 43% of the control cases compared to just 12% of the intervention cases (Ploeg et al. 2009).

The Indian Scenario

Constitutional and Legislative Provisions

Article 4.1 of the Constitution of India speaks about the well-being of the senior citizen as 'The state shall, within the limits of its economic capacity and development, make effective provision for securing the right to public assistance in cases of old age'. A few of the major articles and sections of the Constitution of India regarding the elderly are given below (Piyush and Amir 2016):

Article 41 is reinforced by Section 125 of the code of criminal procedure 1973. In this provision, every person having sufficient means is required to provide for his parents who are unable to maintain themselves.

Section 20(3) of the Hindu Adoption and Maintenance Act 1956 makes it obligatory for a person to maintain his aged or infirm parents.

The Maintenance and Welfare of Parents and Senior Citizens Act, 2007 was landmark legislation piloted by the Ministry of Social Justice and Empowerment. It is being introduced by the states and union territories in stages. Under this act, any senior citizen who is unable to maintain himself/herself from his earning or the earning from any property owned by him is entitled to receive relief. Relatives are also bound under this act to look after senior citizens. In case the relief is not provided by their children, grandchildren or relatives, the elderly can seek assistance from the Tribunal constituted under this act to enforce the remedy of maintenance.

Transfer of property to elderly parents from their children is possible if that property was initially passed to the children on condition they would look after their parent(s).

The Code of Criminal Procedure (CrPC), 1973 is a secular law which governs persons belonging to all religions and communities. Daughters, including married daughters, also have a duty to maintain their parents.

The National Policy on Older Persons 1999 (NPOP) (National Policy on Older Persons 1999) was announced by the Government of India in 1999, which was also the UN International Year on Older Persons (National Policy on Older Persons 1999). It helps to assure the rightful place of the elderly in society so that they can lead a life of dignity. The policy also aims to protect the elderly from exploitation, abuse or neglect. It directs the states to:

- provide financial security
- make provisions so that seniors can be provided with shelter, health-care and other needs
- provide opportunities for the development of seniors' potential
- increase the participation of older people
- provide support services for the elderly
- ensure that the rights of the elderly are not violated
- provide special provisions for elderly females, especially after widowhood.

India as a Signatory to Important International Documents

The Government of India is a signatory to all the following documents (The National Policy of Senior Citizens 2011) and affirms its commitment towards support for older people.

- Proclamation on Ageing and the Global Targets on Ageing for the Year 2001, adopted by the General Assembly in 1992.
- Madrid Plan of Action and the United Nations Principles for Senior Citizens adopted by the UN General Assembly in 2002.
- Shanghai Plan of Action, 2002.
- Macau Outcome Document, 2007, adopted by UNESCAP, which forms the basis for the global policy guidelines to encourage governments to design and implement their own policies from time to time.

The National Policy of Senior Citizens, 2011

In principle, this policy values an age-integrated society (The National Policy of Senior Citizens 2011). It will endeavour to strengthen integration between generations, facilitate interaction between the old and the young as well as strengthen bonds between different age groups. It believes in the development of a formal and informal social support system, so that the capacity of the family to take care of senior citizens is strengthened and they continue to live in the family. The policy seeks to reach out in particular to the bulk of senior citizens living in rural areas who are dependent on family bonds and inter-generational understanding and support.

The National Programme for the Health Care of the Elderly (NPHCE)

NPHCE is an articulation of the government's international and national commitments as envisaged under the UN Convention on the Rights of Persons with Disabilities (UNCRPD), the National Policy on Older Persons (NPOP) adopted in 1999, and Section 20 of the Maintenance and Welfare of Parents and Senior Citizens Act, 2007, dealing with provisions for medical care of senior citizens (National Programme for the Health Care of Elderly (NPHCE) Operational Guidelines 2011).

This programme aims to provide the infrastructure for the delivery of geriatric care and rehabilitation services right from the tertiary level to the PHC level. It also envisages training medical professionals in post-graduate courses in geriatric medicine and other human resources in geriatric care.

Training of Professionals

In line with the NPHCE, medical colleges have been entrusted with offering post-graduate medical degrees in geriatric medicine. They are required to run geriatric clinics and admit patients in geriatric wards specially created for sick elderly patients. Indira Gandhi National Open University is also offering a post-graduate diploma programme in geriatric medicine which aims to orient practising doctors towards the preventive, curative and rehabilitative management of geriatric patients and older persons. A few seminars and lectures are also held by the various hospitals in this area from time to time.

Training of Care-Givers

There are a number of organizations that are involved in the training of care-givers in general. A certificate course in care-giving programmes is offered by the Rehabilitation Council of India (2017). For over a decade, the National Institute of Social Defence (NISD) (2017) has been actively involved in training in the field of social defence including care of older persons. The Old Age Care Division of the Institute runs a series of programmes/certificate courses.

The Institute conducts the courses by itself and in collaboration with regional resource training centres (RRTCs) and other reputed organizations. There are currently four RRTCs in the field of old-age care, designated by the Department of Social Justice and Empowerment. The long-term, short-term and thematic-based courses that are presently being carried out by the NISD/collaborating organizations are a PG diploma in integrated geriatric care, and certificate courses in geriatric care.

Role of NGOs

Various organizations in India have support programmes for the general public or specific groups. The Agewell Foundation (2017) has undertaken a sensitization and training programme for Delhi Police personnel about older people's needs and rights in collaboration with Delhi Police. The Foundation has organized a sensitization and training programme at various police stations in Delhi.

Conclusion

Prevention and management of elder abuse is a complex issue and requires a multi-faceted approach. Commitment must be built at all levels. At the national level, policy, laws and programmes are required. The health-care providers need to be sensitized and trained to handle and report abuse. Society at large and elderly people should be made aware of seniors' rights and recognize the existence of elder abuse and neglect. Seniors can also be involved in designing programmes for spreading awareness and preventing abuse. It is equally important to focus on the research component associated with the abuse of older persons. Ultimately, elder abuse will only be successfully prevented if a culture that nurtures inter-generational solidarity and rejects violence is developed.

References

Action on Elder Abuse, London. 1997. *Hearing the despair: The reality of elder abuse.*

Agewell Foundation. 2017. *Police personnel sensitization and training towards older persons age well foundation.* http://www.agewellfoundation.org/?page_id=4680. Accessed 15 February 2017.

Ansen, P., and R. Breckman. 1988. *Elder mistreatment guidelines: Detection, assessment and intervention.* New York, NY: Mount Sinai/Victim.

Barer, B.M. 1997. The secret shame of the very old: "I've never told this to anyone else". *Journal of Mental Health and Aging* 3: 365–375.

Blumer, H. 1971. Social problems as collective behaviour. *Social Problems* 18: 298–306.

Bomba, P.A. 2006. Use of a single page elder abuse assessment and management tool: A practical clinician's approach to identifying elder mistreatment. *Journal of Gerontological Social Work* 46 (3–4): 103–122.

Bonnie, R.J., and R.B. Wallace. 2003. *Elder mistreatment: Abuse, neglect and exploitation in an aging America.* Washington, DC: National Research Council.

Certificate Course on Care-Giving. *Rehabilitation Council of India.* 2017. http://www.rehabcouncil.nic.in/writereaddata/caregsyll.pdf. Accessed 15 February 2017.

Clark-Daniels, C.L., R.S. Daniels, and L.A. Baumhover. 1990. Abuse and neglect of the elderly: Are emergency staff aware of mandatory reporting laws department? *Annals of Emergency Medicine* 19: 970–977. doi:10.1016/S0196-0644(05)82556-9.

Cooper, C., K. Maxmin, A. Selwood, M. Blanchard, and G. Livingston. 2009. The sensitivity and specificity of the modified conflict tactics scale for detecting clinically significant elder abuse. *International Psychogeriatrics* 21: 774–778.

Dong, X., and M.A. Simon. 2008. Is greater social support a protective factor against elder mistreatment? *Gerontology* 54: 381–388.

Donovan, K., and C. Regehr. 2010. Elder abuse: Clinical, ethical, and legal considerations in social work practice. *Journal of Clinical Social Work* 38: 174–182.

Dumortier, V., and H. de Freminville. 2014. Diagnosis and management of elder abuse by family physician. *Open Journal of Social Sciences* 2: 424–432. doi:10.4236/jss.2014.26049.

Dyer, C.B., C.J. Heisler, C.A. Hill, and L.C. Kim. 2005. Community approaches to elder abuse. *Clinics in Geriatric Medicine* 21 (2): 429–447.

Filinson, R. 1993. An evaluation of a program of volunteer advocates for elder abuse victims. *Journal of Elder Abuse and Neglect* 5 (1): 77–93.

Fulmer, T., G. Paveza, C. Vende Weerd, L. Guadagno, S. Fairchild, R. Norman, A. Abraham, and M. Bolton-Blatt. 2005. Neglect assessment in urban emergency departments and confirmations by an expert clinical team. *Journal of Gerontology*, 60 (8): 1002–1006.

Heath, H., and L. Phair. 2007. Emerging agendas. *Nursing Older People* 19: 16–17.

Hightower, J., M.J. Smith, and H.C. Hightower. 2006. Hearing the voices of abused older women. *Journal of Gerontological Social Work* 46: 205–227.

Hill, J.R. 2005. Teaching about family violence: A proposed model curriculum. *Teach Learn Medicine* 17 (2): 169–178.

Jones, J.S. 1990. Geriatric abuse and neglect. In *Geriatric emergency medicine*, ed. G. Bosker, et al., 533–542. St. Louis, MO: CV Mosby.

Jones, J.S., T.R. Veenstra, and J.P. Seamon. 1997. Elder mistreatment: national survey of emergency physicians. *Annals of Emergency Medicine* 30 (4): 473–479.

Kalaga, H., P. Kingston, B. Penhale, and J. Andrews. 2007. *A review of literature on effective interventions that prevent and respond to harm against adults*. Edinburgh: Scottish Government Social Research.

Keikelame, J., and M. Ferreira. 2000. Mpathekombi, ya bantu abadala: Elder abuse in black townships on the Cape Flats. Cape Town, Human Sciences Research Council and University of Cape Town Centre for Gerontology.

Lachs, M.S., and K.A. Pillemer. 1995. Abuse and neglect of elderly persons. *New England Journal of Medicine* 332: 437–443.

Lachs, M.S., C.S. Williams, S. O'Brien, et al. 1997. ED use by older victims of family violence. *Annals of Emergency Medicine* 30: 448–454. doi:10.1016/S0196-0644(97)70003-9.

Mowlam, A., R. Tennant, J. Dixon, and C. McCreadie. 2007. *UK study of abuse and neglect of older people: Qualitative findings*. London: King's College London and the National Centre for Social Research.

National Institute of Social Defence (NISD). 2017. http://www.nisd.gov.in/content/128_1_OldAgeCare.aspx. Accessed 15 February 2017.

National Policy on Older Persons. 1999. *Ministry of social justice and empowerment*. https://www.scribd.com/document/276944022/National-Policy-on-Older-Persons-1999.

National Programme for the Health Care of Elderly (NPHCE) Operational Guidelines. 2011. http://mohfw.nic.in/showfile.php?lid=1403.

Naughton, C., J. Drennan, M.P. Treacy, A. Lafferty, I. Lyons, A. Phelan, A. O'Loughlin, and L. Delaney. 2010. *Abuse and neglect of older people in Ireland: Report on the national study of elder abuse and neglect*. Dublin: National Centre for the Protection of Older People.

Nerenberg, L. 2002a. *Developing training programs on elder abuse prevention for in-home helpers: Issues and guidelines*. Washington, DC: National Center on Elder Abuse. http://www.ncea.aoa.gov/NCEAroot/Main_Site/pdf/family/training.pdf

Nerenberg, L. 2002b. *Preventing elder abuse by in-home helpers*. Washington, DC: National Center on Elder Abuse. http://www.ncea.aoa.gov/NCEAroot/Main_Site/pdf/publication/preventing.pdf

Nerenberg, L., and D. Baldridge. 2004. *Preventing and responding to abuse of elders in Indian Country*. Washington, DC: National Indian Council on Aging for the National Center on Elder Abuse. No. 90-AP-2144.

Nerenberg, L. 2006. Communities respond to elder abuse. *Journal of Gerontological Social Work* 46: 5–33.

Nerenberg, L. 2008. *Elder abuse prevention: Emerging trends and promising strategies*. New York: Springer.

O'Donnell, D., Margaret P. Treacy, Gerard Fealy, Imogen Lyons, Amanda Phelan, Attracta Lafferty, Jonathan Drennan, Suzanne Quin, and Anne O'Loughlin. 2012. Managing Elder Abuse in Ireland: The Senior Case Worker's Experience Deirdre National Centre for the Protection of Older People (NCPOP) UCD School of Nursing, Midwifery and Heath Systems © UCD and HSE.

Pillemer, K., and D. Finkelhor. 1988. The prevalence of elder abuse: A random sample survey. *The Gerontologist* 28: 51–57.

Pillemer, K.A., K.U. Mueller-Johnson, S.E. Mock, J.J. Suitor, and M.S. Lachs. 2007. Interventions to prevent elder mistreatment. In *Handbook of injury and violence prevention*, ed. L.S. Doll, S. E. Bonzo, J.A. Mercy, and D.A. Sleet, 241–254. New York: Springer.

Piyush, Gupta, and Khan Marrof Amir. 2016. *Chapter on clinical geriatrics in textbook of community medicine*. New Delhi: CBS Publishers and Distributors.

Ploeg, J., J. Fear, B. Hutchison, H. Macmillan, and G. Bolan. 2009. A systematic review of interventions for elder abuse. *Journal of Elder Abuse and Neglect* 21: 187–210.

Podnieks, E. 1999. Support groups: A chance at human connection for abused older adults. In *Elder abuse work*, ed. J. Pritchard, 425–443. London: Jessica Kingsley Publishers.

Podnieks, E. 2006. Social inclusion: An interplay of the determinants of health-new insights into elder abuse. *Journal of Gerontological Social Work* 46: 57–79.

Pritchard, J. 2000. *The needs of older women: Services for victims of elder abuse*. Bristol: The Policy Press.

Regehr, C., and B. Antle. 1997. Coercive influences: Informed consent in court-mandated social work practice. *Social Work* 42 (3): 300–306.

Reingold, D.A. 2006. An elder abuse shelter program: Build it and they will come, a long term care based program to address elder abuse in the community. *Journal of Gerontological Social Work* 46: 123–135.

Rosenblatt, D.E., K.H. Cho, and P.W. Durance. 1996. Reporting mistreatment of older adults: The role of physicians. *Journal of American Geriatric Society* 44 (1): 65–70.

Ross, C. 2007. Detecting and preventing elder abuse. *Nursing and Residential Care* 9: 504.

Ruiz Sanmartín, A., et al. 2001. Violencia doméstica: prevalencia de sospecha de maltrato a ancianos. *Atención Primaria* 27: 331–334.

Shugarman, L.R., B.E. Fries, R.S. Wolf, and J.N. Morris. 2003. Identifying older people at risk of abuse during routine screening practices. *Journal of American Geriatric Society* 51 (1): 24–31.

Soneja, Shubha. 2011. *Elder abuse in India, country report for WHO*. www.who.int/ageing/projects/elder_abuse/alc_ea_ind.pdf.

Srinivasan, Muthuvenkatachalam, and Sandhya Gupta. 2015. The attitude of family towards elder mistreatment: Cultural perspective in rural north India. *Delhi Psychiatry Journal* 18 (1): 119–125.

Stotkowski, L.A. 2008. *Forensic issues for nurses-elder abuse*. http://cme.medscape.com/viewarticle/578859.

Taylor, D.K., G. Bachuwa, J. Evans, and V. Jackson-Johnson. 2006. Assessing barriers to the identification of elder abuse and neglect: A community wide survey of primary care physicians. *Journal of the National Medical Association* 98 (3): 403–404.

The National Policy of Senior Citizens. 2011. http://www.socialjustice.nic.in/writereaddata/UploadFile/dnpsc.pdf.

U.S. Preventative Services Task Force. 2004. Screening for family and intimate partner violence: recommendation statement (review). *Annals of Family Medicine* 2 (2): 156–160.

Wagenaar, D.B., R. Rosenbaum, C. Page, and S. Herman. 2009. Elder abuse education in residency programs: How well are we doing? *Academic Medicine* 84 (5): 611–618.

Wagenaar, Deborah B., Rachel Rosenbaum, Connie Page, and Sandra Herman. 2010. Primary care physicians and elder abuse: Current attitudes and practices. *The Journal of the American Osteopathic Association* 110: 703–711.

Wolfe, D.A. 2003. Elder abuse intervention: Lessons from child abuse and domestic violence initiatives. In *Elder mistreatment: Abuse, neglect and exploitation in an aging America*, ed. R.J. Bonnie, and R.B. Wallace, 501–526. Washington, DC: National Academies Press.

World Health Organization, Geneva. 2002. *Abuse of elderly: World report on violence and health.* http://www.who.int/violence_injury_prevention/violence/global_campaign/en/chap5.pdf. Accessed 15 December 2014.

World Health Organization, Geneva. 2002a. *Abuse of the elderly.* http://www.who.int/violence_injury_prevention/violence/world_report/factsheets/en/elderabusefacts.pdf. Accessed 15 December 2014.

World Health Organization, Geneva. 2002b. Active ageing: A policy framework (WHO/NMH/NPH/02.8).

World Health Organization, Geneva. 2002c. Missing voices: Views of older persons on elder abuse (WHO/NMH/VIP/0.2; WHO/NMH/NPH/02.2).

World Health Organization, Geneva. 2002d. The Toronto declaration on the global prevention of elder abuse.

World Health Organization, Geneva. 2008. *A global response to elder abuse and neglect: Building primary health care capacity to deal with the problem worldwide: Main report.* http://www.who.int/ageing/publications/ELDER_DocAugust08.pdf.

World Health Organization, Geneva. 2015. World report on ageing and health.

Yamada, Y.A. 1999. A Telephone counseling program for elder abuse in Japan. *Journal of Elder Abuse and Neglect* 11: 105–112.

Yan, E., and C.S.K. Tang. 2001. Prevalance and psychological impact of Chinese elder abuse. *Journal of Interpersonal Violence* 16: 1158–1174.

MIX
Papier aus verantwortungsvollen Quellen
Paper from responsible sources
FSC® C105338

If you have any concerns about our products,
you can contact us on
ProductSafety@springernature.com

In case Publisher is established outside the EU,
the EU authorized representative is:
Springer Nature Customer Service Center GmbH
Europaplatz 3, 69115 Heidelberg, Germany

Printed by Libri Plureos GmbH
in Hamburg, Germany